HOW THINGS WORK

AN ILLUSTRATED GUIDE TO THE MECHANICS IN THE WORLD AROUND US

HOW THINGS WORK

AN ILLUSTRATED GUIDE TO THE MECHANICS IN THE WORLD AROUND US

CHARTWELL
BOOKS

Inspiring | Educating | Creating | Entertaining

Brimming with creative inspiration, how-to projects, and useful information to enrich your everyday life, Quarto Knows is a favorite destination for those pursuing their interests and passions. Visit our site and dig deeper with our books into your area of interest: Quarto Creates, Quarto Cooks, Quarto Homes, Quarto Lives, Quarto Drives, Quarto Explores, Quarto Gifts, or Quarto Kids.

10 9 8 7 6 5 4 3 2 1

ISBN: 978-0-7858-3888-3

Original idea Joan Ricart
Editorial coordination Emilio López
Design Clara Miralles, Ana Roca
Editing Alberto Hernández
Layout Clara Miralles, Carla Cobas

Printed in Singapore COS062020

PREFACE

With full-color cross sections, *How Things Work*, finally simplifies the mechanics of the world around us. More than 100 things are dissected so that one can examine the inner workings of things as diverse as a 3-D printer or a television.

Learn about how these things developed over time and how they impacted the course of human development. From ancient chariots of war, to the telegraph, to the technologies of the future, learn about the mechanics of the world around us.

The book is organized categorically into ten chapters, covering topics like Transportation, Architecture, Communication, and Ancient Civilizations. Each chapter has eleven subjects that are dissected through diagrams and cross-sections. A thematic index at the end allows one to easily locate all items of interest mentioned in these pages.

Contents

TRANSPORTATION

Balloons

Balloons were the first successful flight technique developed by humans. Although the first recorded manned balloon flight was carried out by the Montgolfier brothers of France, the Chinese used unmanned balloons for military communication in the 2nd century AD. Because balloons are carried along by the wind, the pilot often has a difficult time following an exact course or returning to the place of origin. Forgotten by the beginning of the 20th century, balloons have experienced a revival since the 1960s and are now used for sport and recreation.

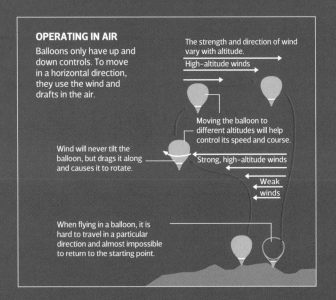

OPERATING IN AIR

Balloons only have up and down controls. To move in a horizontal direction, they use the wind and drafts in the air.

The strength and direction of wind vary with altitude.

High-altitude winds

Moving the balloon to different altitudes will help control its speed and course.

Wind will never tilt the balloon, but drags it along and causes it to rotate.

Strong, high-altitude winds

Weak winds

When flying in a balloon, it is hard to travel in a particular direction and almost impossible to return to the starting point.

Why they fly?

Hot-air balloons fly thanks to differences in temperature. Some gases and hot air are lighter than atmospheric air.

CONTROL VALVE

In bigger balloons, the control valve is used to control altitude.

Closed
When closed, hot air remains inside the balloon forcing it to rise.

Opened
When the valve is opened, some of the hot air escapes and the balloon descends.

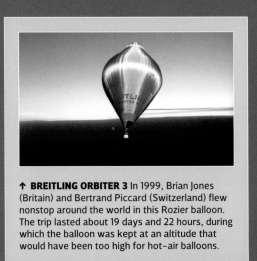

↑ **BREITLING ORBITER 3** In 1999, Brian Jones (Britain) and Bertrand Piccard (Switzerland) flew nonstop around the world in this Rozier balloon. The trip lasted about 19 days and 22 hours, during which the balloon was kept at an altitude that would have been too high for hot-air balloons.

When the balloon is filled with air that is less dense than the atmospheric air surrounding it, the balloon rises.

A propane gas burner heats up the air inside the balloon. As the air molecules are heated, the air expands and becomes less dense.

HOT-AIR BALLOONS

A propane gas burner heats the air inside the balloon.

As the air heats up, it expands and becomes less dense. Cold air is heavier and tends to descend. Hot air is lighter and rises.

GAS BALLOONS

HELIUM

Are mostly used in unmanned meteorological missions. They are usually filled with hydrogen or helium, both light gases.

ROZIER BALLOONS

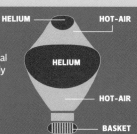

HELIUM — HOT-AIR

HELIUM

HOT-AIR

BASKET

Are a combination of helium and hot-air balloons. They allow for longer journeys and higher altitudes.

Segments
There are balloons with 8, 16, or even 24 segments.

Rip panels

PRESSURE, WEIGHT AND ALTITUDE

Gases are found at different altitudes:

Less weight and more pressure: More altitude.

More weight and less pressure: Less altitude.

AIR

EARTH

1783

The Montgolfier brothers of France fly for the first time in a hot-air balloon.

The bottom part is open to allow the burner to heat the air.

Skirt
Some balloons have a skirt made of nonflammable material that prevents the nylon from catching fire during the inflation process.

Envelope
Made of nylon or polyester coated in durable and low-weight polyurethane (to keep air from escaping), it contains the heated air.

Burners
Use propane, just like the portable stoves used in camping.

Gondola
For the pilot and passangers.

The airship

There was a time when it was possible to cross the ocean floating in the air, surrounded by luxury and fine dining, resting at night in comfortable cabins, enjoying every inch of the landscape through huge windows above the sea. It was the time of airships — some of which were the largest flying machines ever built. Airships connected America and Europe through the sky, before the age of airplanes. Its golden days were short and ended when aeroplanes — faster, safer and cheaper — gained popularity by the end of 1930's. However, its glamour and mystique have not yet disappeared.

The Hindenburg

It was the biggest airship, characterized by the LZ-129, built by the Zeppelin shipyard. In 1937, when landing in New Jersey it caught fire and blew up. Since then, no airships have been used for trading flights. ↘

The structure
It was rigid, made of an aluminum covering with a big cotton tissue.

Gas vent

Nose/anchor
It allows the airship to moor at special docks.

← FERDINAND VON ZEPPELIN
German aeronaut. In 1873, he had a dream of being able to fly in his own airship. In 1900 he fulfilled his dream. Ten years later his airships started to transport passengers.

Directional aerials
Used to focus radiation on specific spots.

Control cabin
A crew of 9 flew the airship.

HOW THE AIRSHIPS FLEW

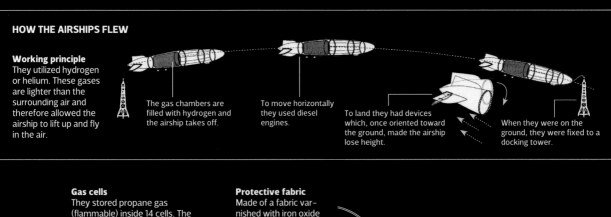

Working principle
They utilized hydrogen or helium. These gases are lighter than the surrounding air and therefore allowed the airship to lift up and fly in the air.

The gas chambers are filled with hydrogen and the airship takes off.

To move horizontally they used diesel engines.

To land they had devices which, once oriented toward the ground, made the airship lose height.

When they were on the ground, they were fixed to a docking tower.

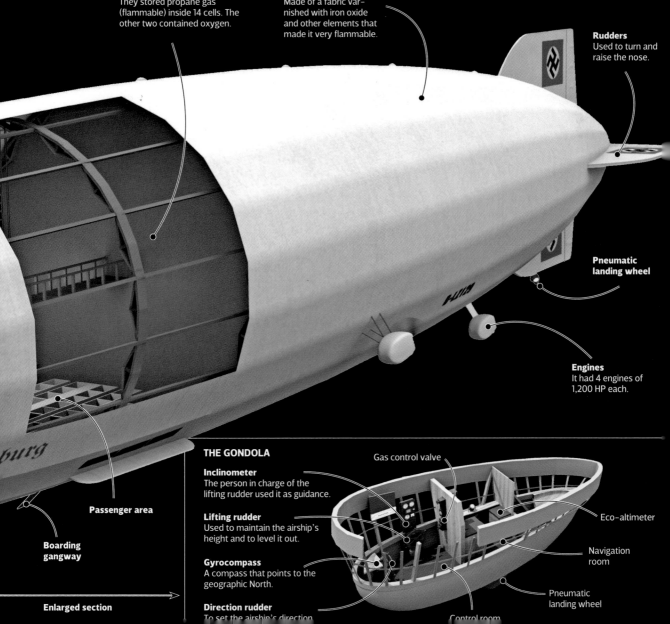

Gas cells
They stored propane gas (flammable) inside 14 cells. The other two contained oxygen.

Protective fabric
Made of a fabric varnished with iron oxide and other elements that made it very flammable.

Rudders
Used to turn and raise the nose.

Pneumatic landing wheel

Engines
It had 4 engines of 1,200 HP each.

Passenger area

Boarding gangway

Enlarged section

THE GONDOLA

Inclinometer
The person in charge of the lifting rudder used it as guidance.

Lifting rudder
Used to maintain the airship's height and to level it out.

Gyrocompass
A compass that points to the geographic North.

Direction rudder
To set the airship's direction

Gas control valve

Eco-altimeter

Navigation room

Pneumatic landing wheel

Control room

Airplanes

Airplanes forever changed passenger transportation around the world forever and made possible what, for centuries, had seemed like an impossible dream: crossing the planet in hours. By the mid-20th century, bigger airplanes already transported thousands of people at the same time in a comfortable and safe way at a relatively low cost. Airplanes are the icon of the modern globalized world. It is difficult to calculate the number of flights taken around the world every day, but the number of passengers transported is high and ever-increasing.

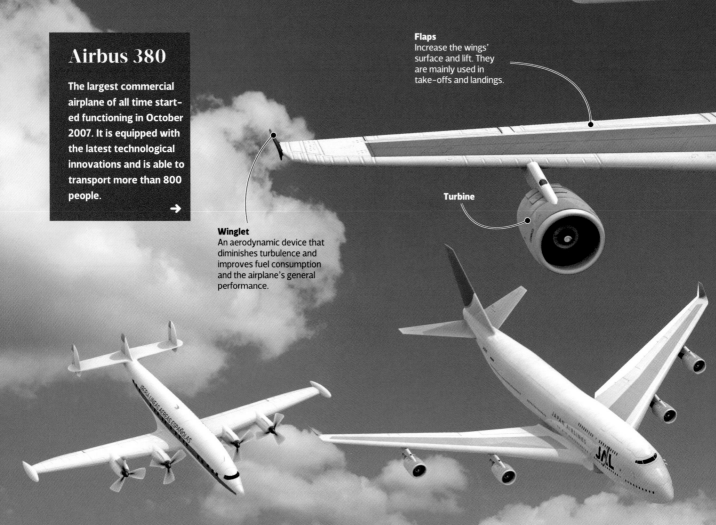

Airbus 380

The largest commercial airplane of all time started functioning in October 2007. It is equipped with the latest technological innovations and is able to transport more than 800 people.

→

Flaps
Increase the wings' surface and lift. They are mainly used in take-offs and landings.

Turbine

Winglet
An aerodynamic device that diminishes turbulence and improves fuel consumption and the airplane's general performance.

↑ **SUPER CONSTELLATION** The appearance of the DC-3 in the 1930's, with its high safety standards, its new navigational aids, and its nearly unprecedented focus on passenger comfort turned it into a milestone of air transportation history.

↑ **BOEING 747 "JUMBO"** The Boeing 747 was the unquestioned king of the skies for almost four decades—until the emergence of the Airbus A-380. Able to comfortably seat 524 passengers, the 747 was the icon of the Jet Age.

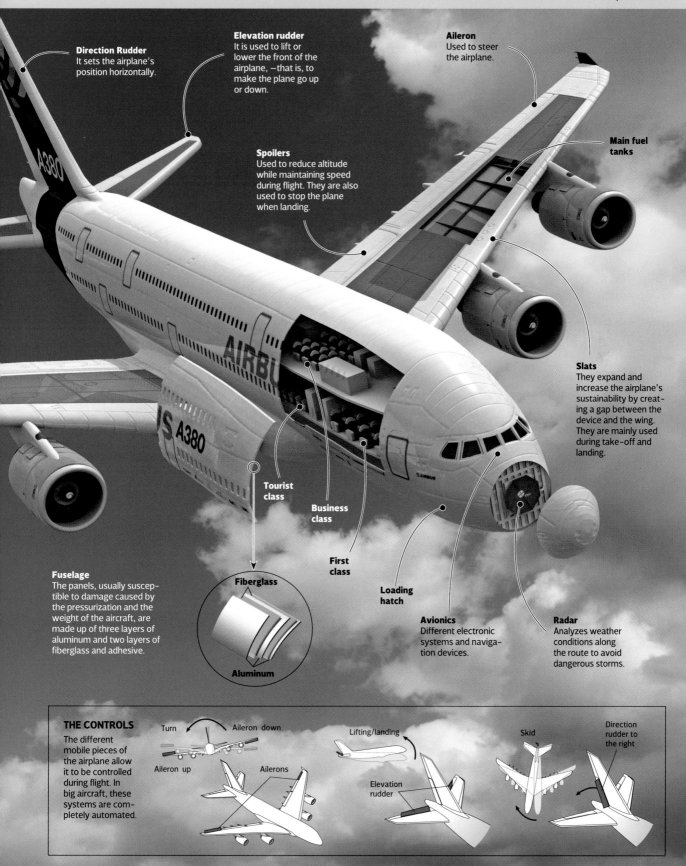

Direction Rudder
It sets the airplane's position horizontally.

Elevation rudder
It is used to lift or lower the front of the airplane, —that is, to make the plane go up or down.

Aileron
Used to steer the airplane.

Main fuel tanks

Spoilers
Used to reduce altitude while maintaining speed during flight. They are also used to stop the plane when landing.

Slats
They expand and increase the airplane's sustainability by creating a gap between the device and the wing. They are mainly used during take-off and landing.

Tourist class

Business class

First class

Loading hatch

Avionics
Different electronic systems and navigation devices.

Radar
Analyzes weather conditions along the route to avoid dangerous storms.

Fuselage
The panels, usually susceptible to damage caused by the pressurization and the weight of the aircraft, are made up of three layers of aluminum and two layers of fiberglass and adhesive.

Fiberglass

Aluminum

THE CONTROLS
The different mobile pieces of the airplane allow it to be controlled during flight. In big aircraft, these systems are completely automated.

Turn
Aileron down
Aileron up
Ailerons

Lifting/landing
Elevation rudder

Skid

Direction rudder to the right

The helicopter

The history of the helicopter dates back to the 20th century, though it did not became popular until the 1940's. Ubiquitous in skies all over the world, the helicopter has never been a means of mass transportation, however, it has found the best results in the military field, fire fighting, and search and rescue operations. They provide passenger services in places that are difficult to access using other vehicles, though they are limited by their small seating.

Dexterities

Despite its limited transportation load, helicopters can perform dexterous maneuvers. They require very little space to land or take-off and can remain suspended in the air over a specific spot. Some can even fly backwards.

→

Tail structure

Tail fin

Tail rotor
The turning of the propellers causes the body of the helicopter to turn in the opposite direction. To avoid this, a tail rotor is employed.

Blades
They have an aerodynamic shape similar to the wings of a plane. Their angles can change and produce varying intensities of lift to service different types of flight.

Tail stabilizer

DIFFERENT CONFIGURATIONS

Traditional helicopters have a main propeller and a tail rotor. But there are other configurations.

↑ **TANDEM ROTORS** They compensate the turn with the use of two propellers turning in opposite directions. They are often used as transportation vehicles due to their great lift capacity.

↑ **COAXIAL ROTORS** The blades turn in opposite directions on a single shaft. They have a great lift capacity. They are less noisy than traditional helicopters and can operate in more compact spaces.

↑ **NOTAR SYSTEM** Instead of a rotor it uses an air jet to compensate the turning tendency of the helicopter's body. In this way, it improves safety and is one of the less noisy.

WHY DOES IT FLY?

The same lifting system used in planes (related to the shape of the wings) is used in helicopters. The helicopter gets lift by turning its propellers.

Normal position

Better lift

Less lift

The edge of the propellers is the same as a wing. When turning, that is, when producing a flow of air that crosses them, they create lift and the helicopter takes-off.

Chord

Wind

Attack angle

← **PIONEER EXPERIENCE**
Despite an autogyro machine being one of Leonardo da Vinci's ideas in the fifteenth century, it would be 1919 before an Argentinian lawyer and engineer Raúl Pateras Pescara developed a working prototype of the helicopter.

Engine exhaust

Main rotor

Oscillating plate
This is an essential piece of the helicopter because it allows the pilot to modify the inclination of the set of blades allowing the craft to move forward or to remain stationary.

Passenger cabin door

Passenger seats

Landing gear
Landing skids can be used to land on snow or ice.

Cabin door

Crew seats

Front undercarriage

Instrument panel

Fuselage

Tubo Pitot
A sensor which records atmospheric pressure and registers elevation and horizontal/vertical velocities.

The automobile

Transportation around the world relies on the automobile—an invention that is only 100 years old. Although technological advances have increased efficiency and passenger comfort, the operating principle and driving techniques have changed very little since 1885, when Karl Benz presented his tricycle powered by an internal combustion engine. About 600 to 800 million cars provide transportation on roads around the world, showing the extent that the automobile has modified society and penetrated culture.

A complex combination

Present day automobiles combine different systems: mechanic, electric and hydraulic. The engine turns chemical energy into movement. The steering and the brake systems allow controlling the vehicle.

FOUR STROKE ENGINE

Most modern automobiles use this kind of engine to generate traction from air, fuel, and an electrical spark.

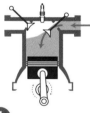

 The piston goes down. By doing so, air and fuel enter the cylinder through the admission valve that is open.

 The piston rises. The admission valve is closed and air is compressed.

 A spark produced by the spark plug ignites the mixture. The expansion of the gases forces the piston to go down creating movement.

Valve stem   Spark plug   Valve stem

Piston — — Con-necting rod

The piston rises. Residual gases from the combustion are eliminated through the escape valve that remains open.

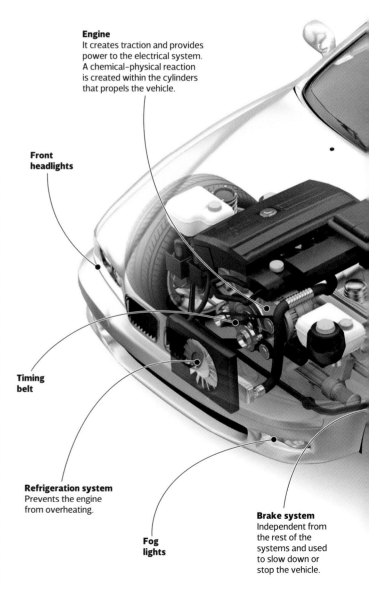

Engine
It creates traction and provides power to the electrical system. A chemical–physical reaction is created within the cylinders that propels the vehicle.

Front headlights

Timing belt

Refrigeration system
Prevents the engine from overheating.

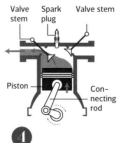

Fog lights

Brake system
Independent from the rest of the systems and used to slow down or stop the vehicle.

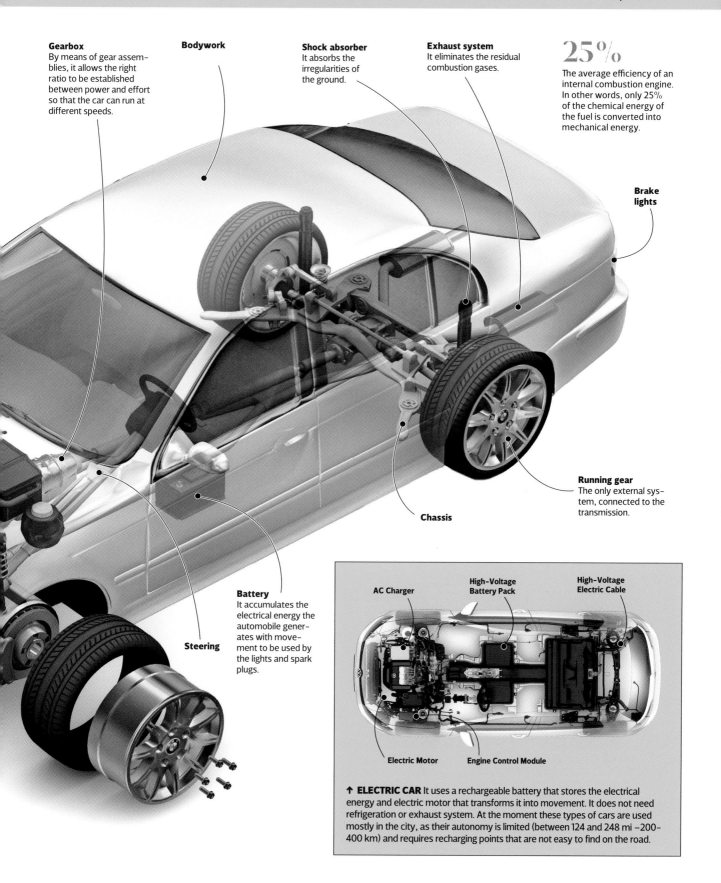

Gearbox
By means of gear assemblies, it allows the right ratio to be established between power and effort so that the car can run at different speeds.

Bodywork

Shock absorber
It absorbs the irregularities of the ground.

Exhaust system
It eliminates the residual combustion gases.

25%
The average efficiency of an internal combustion engine. In other words, only 25% of the chemical energy of the fuel is converted into mechanical energy.

Brake lights

Running gear
The only external system, connected to the transmission.

Chassis

Battery
It accumulates the electrical energy the automobile generates with movement to be used by the lights and spark plugs.

Steering

AC Charger

High-Voltage Battery Pack

High-Voltage Electric Cable

Electric Motor

Engine Control Module

↑ **ELECTRIC CAR** It uses a rechargeable battery that stores the electrical energy and electric motor that transforms it into movement. It does not need refrigeration or exhaust system. At the moment these types of cars are used mostly in the city, as their autonomy is limited (between 124 and 248 mi –200-400 km) and requires recharging points that are not easy to find on the road.

High speed trains

Operating a train within the security standards above 186 mph (300 km/h) requires not only more powerful engines and an aerodynamic design, but a complete redesign of the tracks. In addition, high speed train systems require that the tracks be completely isolated by means of wire fences or embankments to avoid the intrusion of people or animals. China, with a network of more than 6,800 miles (11,000 kilometers) of tracks and more than 3,700 miles (6,000 kilometers) under construction is the leading country in the world for developing high speed trains. Japan, Spain, and France follow it.

France in the lead

The TGV (High Speed Train) is one of the most widespread high speed train systems of the world. Developed in France, it reaches speeds over 186 mph (300 km/h). The TGV only works on electrified rails. ↘

357.16 mph
(574.8 km/h)

Is the speed record achieved by a TGV train on 3 April 2007 on the Paris–Strasbourg line. However, passenger services do not exceed 198 mph (320 km/h).

Cold air sinks

Wheel brake control

Crash protector
It is a "beehive" structure that, when distorted, is able to partially absorb the energy of a potential impact.

ONE TRUCK, TWO CARRIAGES

Unlike traditional trains where the cars have their own trucks, high speed train wagons share the trucks. This system reduces the probability that, in case of the derailment of one wagon, the rest of the train does not overturn.

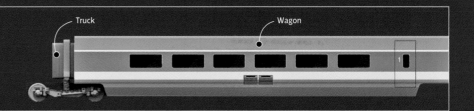

Truck

Wagon

Pantograph for 25,000 V
It unfolds to take the power from the high voltage cables along special high speed train lines.

Pantograph for 1,500 V
It is used in older lines with lower voltage.

Traction engines

Isolation for high voltage cables
A cable transports the power from the pantographs to the engines across the train's roof.

Main compressor
It compresses the air used by the brake system. It normally uses air compressed to 8 or 9 times the average atmospheric pressure.

Toolbox

Battery boxes

Main transformer
It adapts the energy taken from the high voltage wires to the one needed by the train's engines (from 25,000 V to 1,500 V).

Air conditioning system for the driver's cabin.

↑ **TGV** High speed train at Avignon TVG station, in France.

The bicycle

This two–wheel vehicle is not only a healthy, environmentally friendly, and economical means of transportation, it is also extraordinarily efficient! Up to 99 percent of the energy a cyclist transfers to the pedals reaches the wheels. In fact, it is the most efficient load–bearing vehicle. Bicycles have newfound importance in the age of environmental awareness and healthy living because it brings the two together. This icon of clean transportation is being revolutionized in the 21st century by helping to solve transport problems in big cities.

Seat or saddle

Top tube

Seat post

Rear brakes

Seat tube

Seat stay

Chain stay

A great idea

Created in the 19th century, it envolved fast, although it maintains the basic ideas and designs of the earliest bicycles. →

Sprocket wheel
Bicycles can have more than one. They receive the cyclist's power that is transmitted through the chain to the wheels.

Spokes
They connect the rim to the hub, adding structural rigidity to the wheel with only a negligible addition of weight.

Rear derailleur
Maintains tension in the chain.

Chain
It connects the plate with the pinion to activate the rear wheel.

Crank

Pedal

↓ **FAST EVOLUTION** The bicycle achieved perfection early, with only minor improvements to the first designs leading to today's variety of bicycles.

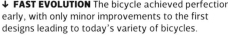

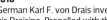

1818
German Karl F. von Drais invents his Draisine. Propelled with the feet.

1839
Kirkpatrick Macmillan adds levers to the Draisine, to move the rear wheel.

1860
Blacksmith Pierre Michaux adds pedals to the front wheels.

1870
The high wheel bicycle, or penny–farthing is created. It permitted a higher speed by pedaling less but was unstable.

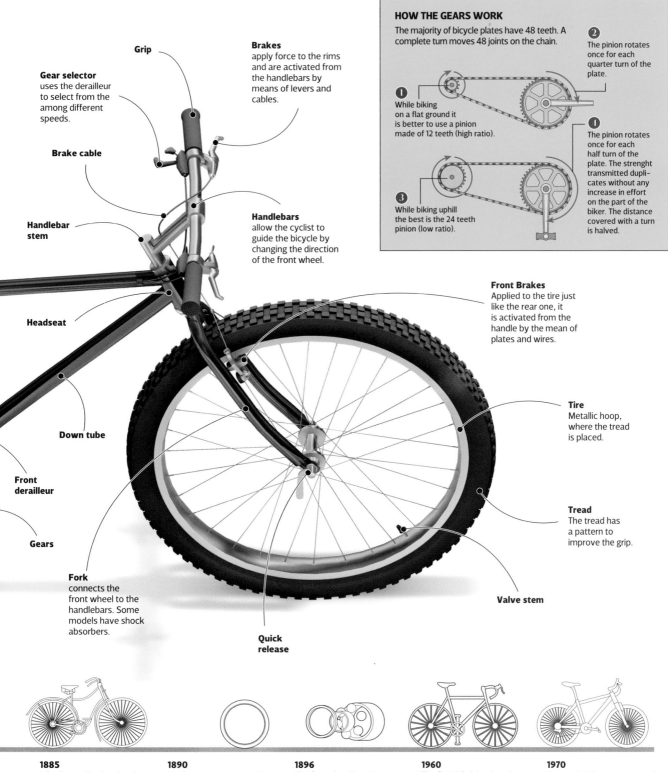

Gear selector
uses the derailleur to select from the among different speeds.

Grip

Brakes
apply force to the rims and are activated from the handlebars by means of levers and cables.

Brake cable

Handlebar stem

Headseat

Handlebars
allow the cyclist to guide the bicycle by changing the direction of the front wheel.

Down tube

Front derailleur

Gears

Fork
connects the front wheel to the handlebars. Some models have shock absorbers.

Quick release

Front Brakes
Applied to the tire just like the rear one, it is activated from the handle by the mean of plates and wires.

Tire
Metallic hoop, where the tread is placed.

Tread
The tread has a pattern to improve the grip.

Valve stem

HOW THE GEARS WORK

The majority of bicycle plates have 48 teeth. A complete turn moves 48 joints on the chain.

① While biking on a flat ground it is better to use a pinion made of 12 teeth (high ratio).

③ While biking uphill the best is the 24 teeth pinion (low ratio).

② The pinion rotates once for each quarter turn of the plate.

① The pinion rotates once for each half turn of the plate. The strenght transmitted duplicates without any increase in effort on the part of the biker. The distance covered with a turn is halved.

1885
John Kemp Starley develops a bicycle with pedals, chain, brakes and same-sized wheels.

1890
Dunlop invents pneumatic inner tube tires and uses them in bicycles.

1896
The freewheel bearing is invented, allowing the rider to stop pedalling while the wheel keeps on turning.

1960
The first ride bicycle arrives on the market.

1970
Mountain bikes are created.

The ship

Although airplanes have replaced ships as the main method of moving people across the world's oceans, they remain very important in moving people across short distances. Today thousands of ferries and ships of all sizes constantly sail across rivers and seas. The big transatlantic liners, which had their golden age before the Second World War, have evolved into huge luxurious floating hotels dedicated to tourism and enjoyment: modern cruise ships.

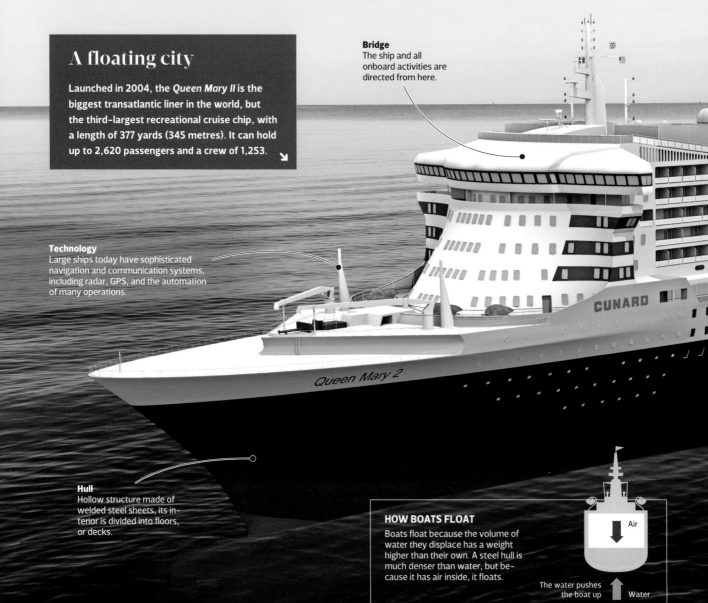

A floating city

Launched in 2004, the *Queen Mary II* is the biggest transatlantic liner in the world, but the third-largest recreational cruise chip, with a length of 377 yards (345 metres). It can hold up to 2,620 passengers and a crew of 1,253.

Bridge
The ship and all onboard activities are directed from here.

Technology
Large ships today have sophisticated navigation and communication systems, including radar, GPS, and the automation of many operations.

CUNARD

Queen Mary 2

Hull
Hollow structure made of welded steel sheets, its interior is divided into floors, or decks.

HOW BOATS FLOAT
Boats float because the volume of water they displace has a weight higher than their own. A steel hull is much denser than water, but because it has air inside, it floats.

Air

The water pushes the boat up   Water

HOW TO MOVE BOATS

Moving efficiently over the water has been a constant concern for sailors. Here is a summary of the main methods.

Oar
One of the most primitive ways to move ships, oars rely on people to provide energy.

Sail
Another primitive way to drive ships, sails take advantage of wind's force. Experienced sailors can guide ships against the direction of the wind.

Paddlewheel
The first ships with engines used steam power to move a paddlewheel. The paddlewheel is always half below the water and half above it.

Propellers
Modern ships use propellers that are completely submerged and are much more efficient than paddlewheels.

Enjoyment
It has golf, tennis, and basketball courts, a 25,026 ft² (2,325 m²) gymnasium, 5 swimming pools, a museum, night club, casino, 14 bars, a cinema theatre, theatre, and an auditorium.

1,187 ft (362 m)
The biggest passenger ship in the world is a cruise ship that is 1,187 ft (362 m) long and 197 ft (60 m) wide. It's name is the *Allure of the Seas* and it can hold up to 6,400 passengers.

Waterline

Cabins
Out of a total of 1,310 cabins, 920 have ocean views, 293 have interior views, and 97 are suites. Cabins range in size from 59 ft² (18 m²) to 686 ft² (209 m²).

Lifeboats
They are located 82 ft (25 m) above the waterline to avoid being damaged by the North Sea's waves.

THE PROPELLER

① The propeller spins due to the power provided by the engines.

Blade

② The water is driven backwards.

③ The reactionary force pushes the boat forwards.

Sailing ships

It is impossible to determine who first thought of taking advantage of the wind to propel a boat over water. However, this discovery accelerated humanity's exploration of the globe. Society sailed into the modern age. The sailing ship was a huge advance over earlier boats that used oars and human power for movement because it relied on non–human energy. In addition to exploratory travel, the sailing ship was used in trade and war until steamboats became popular. Steam engines have now been cast aside in favor of electric engines.

The caravel

This typical sailing ship was the most common in the Atlantic Ocean in the 15th and 17th centuries. The Spanish used the caravel to discover a new world beyond the known realm in 1492. In a sense, caravels and ships allowed the shift from cabotage navigation port-to-port along the coast, to deep sea navigation.

→

HOW THEY NAVIGATE

Sailboats receive different forces, especially from the wind. And the combination of sails and rudder can generate a resultant force with a specific direction towards which the boat will sail.

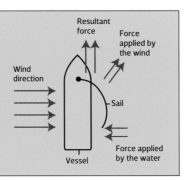

Resultant force

Force applied by the wind

Wind direction

Sail

Force applied by the water

Vessel

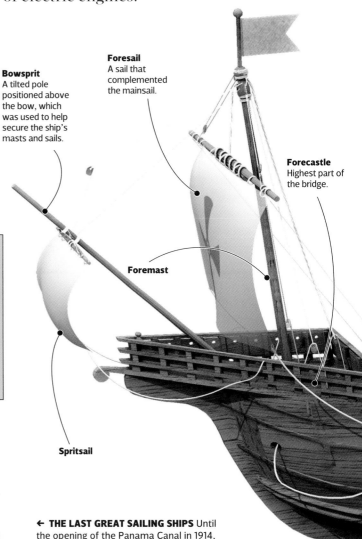

Bowsprit
A tilted pole positioned above the bow, which was used to help secure the ship's masts and sails.

Foresail
A sail that complemented the mainsail.

Forecastle
Highest part of the bridge.

Foremast

Spritsail

← **THE LAST GREAT SAILING SHIPS** Until the opening of the Panama Canal in 1914, Cape Horn, at the tip of South America, was a key maritime trade routes. To overcome its harsh sailing conditions, European shipping companies built huge sailboats of great tonnage and up to six masts.

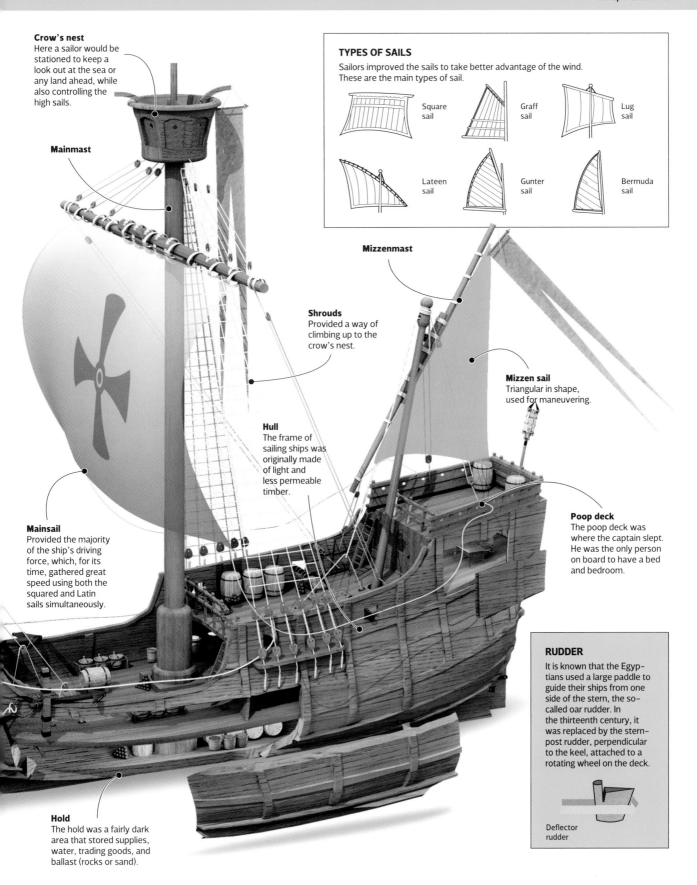

Crow's nest
Here a sailor would be stationed to keep a look out at the sea or any land ahead, while also controlling the high sails.

Mainmast

TYPES OF SAILS
Sailors improved the sails to take better advantage of the wind. These are the main types of sail.

Square sail

Graff sail

Lug sail

Lateen sail

Gunter sail

Bermuda sail

Mizzenmast

Shrouds
Provided a way of climbing up to the crow's nest.

Mizzen sail
Triangular in shape, used for maneuvering.

Hull
The frame of sailing ships was originally made of light and less permeable timber.

Poop deck
The poop deck was where the captain slept. He was the only person on board to have a bed and bedroom.

Mainsail
Provided the majority of the ship's driving force, which, for its time, gathered great speed using both the squared and Latin sails simultaneously.

RUDDER
It is known that the Egyptians used a large paddle to guide their ships from one side of the stern, the so-called oar rudder. In the thirteenth century, it was replaced by the sternpost rudder, perpendicular to the keel, attached to a rotating wheel on the deck.

Deflector rudder

Hold
The hold was a fairly dark area that stored supplies, water, trading goods, and ballast (rocks or sand).

Hovercraft

In the 1960s it seemed like every possible method of passenger transportation by water had already been invented. However, Great Britain entered the scene with the hovercraft, an ingenious solution for crossing the English Channel in about two hours, floating over the waves on an air cushion. Hovercraft had their glory days fifty years ago although they still transport people over difficult terrain. They are also able to cross all irregular terrain, whether solid, liquid, or swampy.

Almost suspended

The powerful fans of a hovercraft keep the vehicle floating, which is why, even though it looks like a ship, it is considered a type of aircraft. ↘

Cabin
Thought they had a lot of space to carry passengers, passengers were not well-insulated from the noisy fans.

Radar
It detects obstacles in the route.

Command cabin

Skirt
It is flexible and prevents air from escaping, which maintains the cushion effect.

↑ **MODERN** While no longer used to cross the English Channel, hovercraft are used to cross difficult terrain, including swampy regions of Alaska, India and Norway. In addition, it is used by militaries and coast guards.

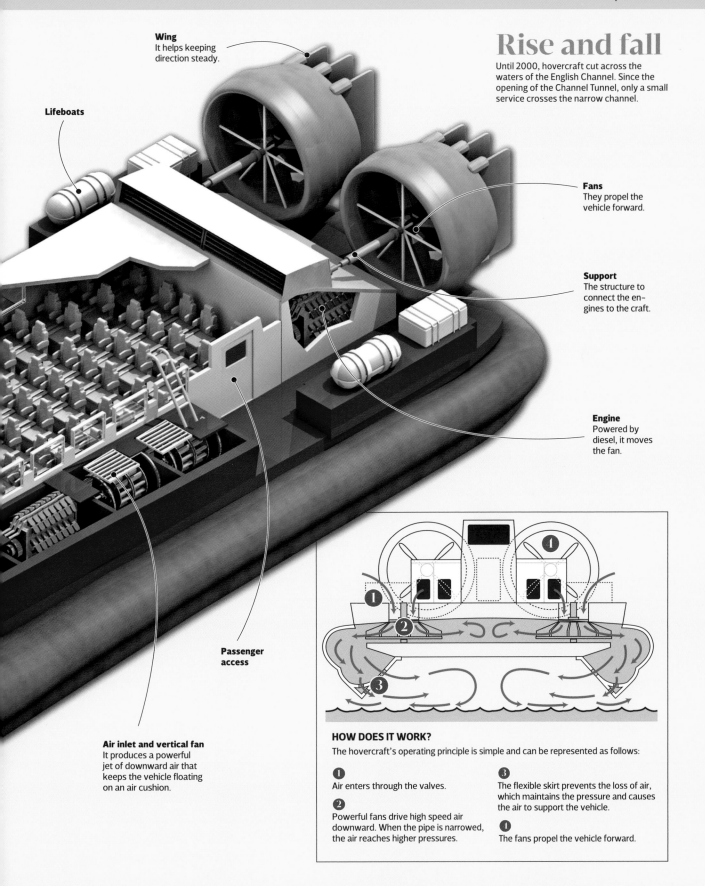

Wing
It helps keeping direction steady.

Lifeboats

Rise and fall

Until 2000, hovercraft cut across the waters of the English Channel. Since the opening of the Channel Tunnel, only a small service crosses the narrow channel.

Fans
They propel the vehicle forward.

Support
The structure to connect the engines to the craft.

Engine
Powered by diesel, it moves the fan.

Passenger access

Air inlet and vertical fan
It produces a powerful jet of downward air that keeps the vehicle floating on an air cushion.

HOW DOES IT WORK?

The hovercraft's operating principle is simple and can be represented as follows:

1
Air enters through the valves.

2
Powerful fans drive high speed air downward. When the pipe is narrowed, the air reaches higher pressures.

3
The flexible skirt prevents the loss of air, which maintains the pressure and causes the air to support the vehicle.

4
The fans propel the vehicle forward.

The cable car

Cable transportation makes up for its slow speed and operational difficulty by having a unique capability: there is neither terrain steep enough nor obstacles difficult enough that the cable railway can't overcome. This makes them essential links in some mass transportation systems. An essential piece of transportation in mountainous landscapes, some of them are famous for their history and the beauty of the landscape they cross. They offer unique trips that turned them into tourist icons.

Hanging from a wire

Despite the high safety standards, the feeling of traveling in a cabin, literally hanging from a thread, generates great fear in people. But, what does a cable car consist of?

Cables
The soul of the cable railway. They provide traction and support to the cabins. They are usually very strong and made of steel.

Stations
They include both the ascending and the descending passenger stations, in addition to the traction engines that operate the whole system.

Towers
They provide power and support to cable cars. Some cable cars have them, while others don't. They usually reach a height of approximately 131 ft (40 m).

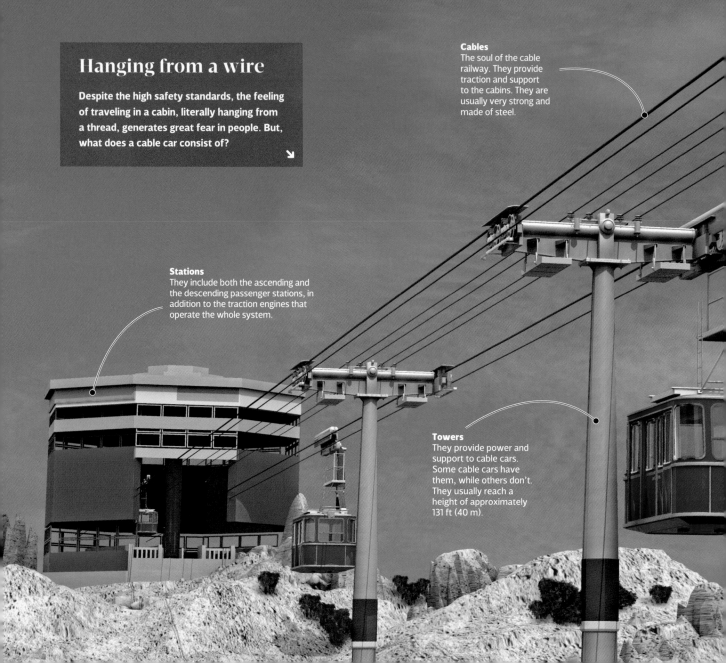

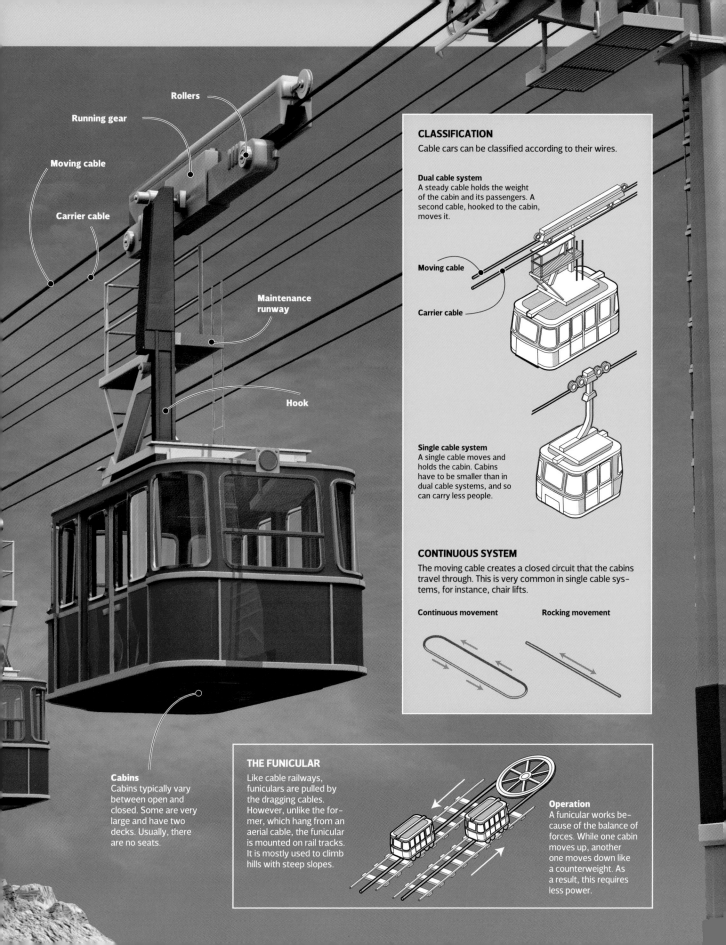

Rollers

Running gear

Moving cable

Carrier cable

Maintenance
runway

Hook

CLASSIFICATION

Cable cars can be classified according to their wires.

Dual cable system

A steady cable holds the weight of the cabin and its passengers. A second cable, hooked to the cabin, moves it.

Moving cable

Carrier cable

Single cable system

A single cable moves and holds the cabin. Cabins have to be smaller than in dual cable systems, and so can carry less people.

CONTINUOUS SYSTEM

The moving cable creates a closed circuit that the cabins travel through. This is very common in single cable systems, for instance, chair lifts.

Continuous movement

Rocking movement

Cabins

Cabins typically vary between open and closed. Some are very large and have two decks. Usually, there are no seats.

THE FUNICULAR

Like cable railways, funiculars are pulled by the dragging cables. However, unlike the former, which hang from an aerial cable, the funicular is mounted on rail tracks. It is mostly used to climb hills with steep slopes.

Operation

A funicular works because of the balance of forces. While one cabin moves up, another one moves down like a counterweight. As a result, this requires less power.

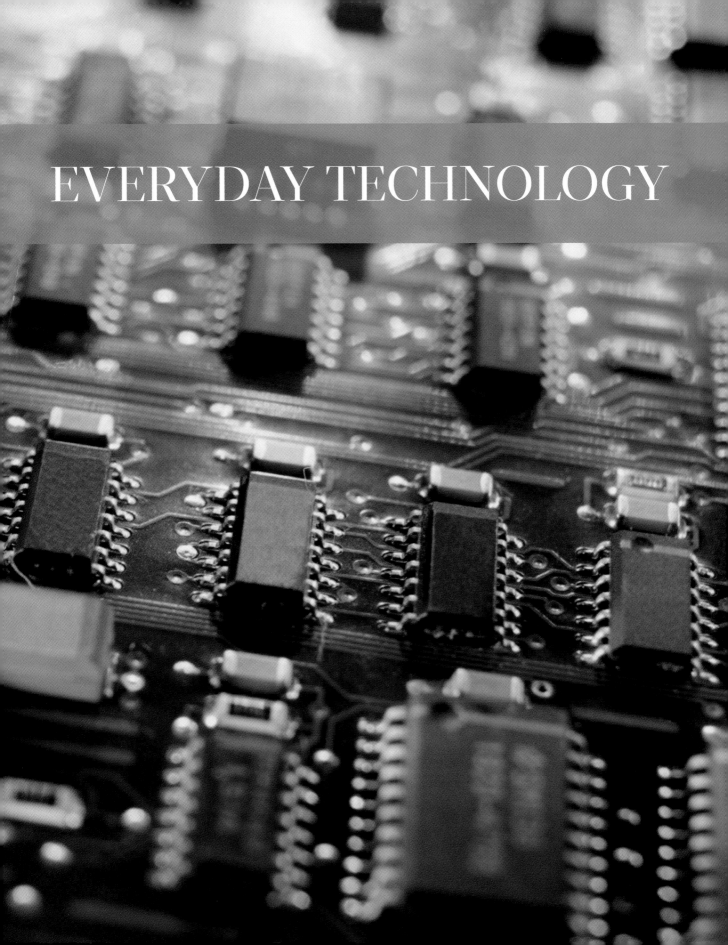

EVERYDAY TECHNOLOGY

The microchip

Tiny as it is, the microchip (or integrated circuit) is the brain of computerized systems, the intelligence that makes all of a computer's components function in a coordinated way. The earliest microchip appeared almost 40 years ago, and since then its capabilities have exponentially increased while its components have shrunk to a microscopic size. Specialists are currently working to develop devices on a molecular scale that will take computerized potential to levels that are currently unimaginable.

↑ THE SMALLEST BRAIN IN THE WORLD Millions of components forming the most complex integrated circuits developed by humanity are assembled in a space of just a few tenths of a square inch. Microprocessors work on the basis of "logic gates" in a "language" written in long sequences of two numbers: one and zero.

6,000

the number of calculations per second that could be handled by the Intel 4004 processor, considered to be the first microprocessor.

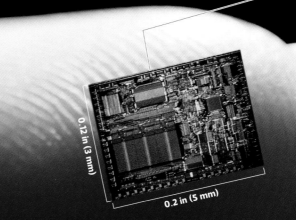

0.12 in (3 mm)

0.2 in (5 mm)

FUNCTIONING OF THE TRANSISTORS

Etched in the silicon, the transistor is a very effective semiconductor device and amplifier, although it is microscopic in size. It acts like an electronic switch with the capacity to activate and deactivate itself by means of an electrical signal.

ACTIVE CIRCUIT

Negative silicon has atoms with free electrons.

Electrical signal

Negative silicon

Positive silicon has atoms without free electrons.

Positive silicon receives free electrons from the electrical signal, establishing a flow of current between the areas of negative silicon, thereby activating the circuit.

INACTIVE CIRCUIT

The current ceases to flow and the circuit is deactivated.

The electrical signal is interrupted.

Miniature circuits

This very thin chip contains an enormous number (in the range of billions) of interconnected micro-electronic devices, mainly diodes and transistors, as well as passive components such as resistors and condensers.

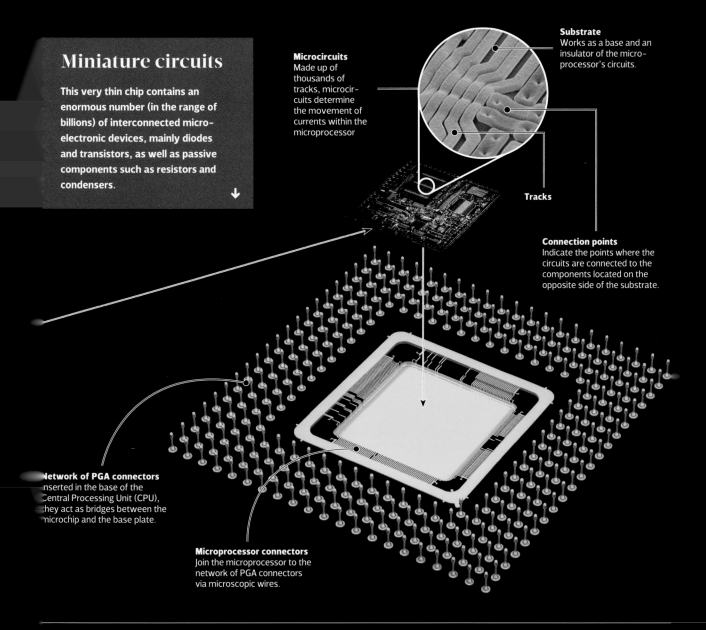

Microcircuits
Made up of thousands of tracks, microcir-cuits determine the movement of currents within the microprocessor

Substrate
Works as a base and an insulator of the micro-processor's circuits.

Tracks

Connection points
Indicate the points where the circuits are connected to the components located on the opposite side of the substrate.

Network of PGA connectors
Inserted in the base of the Central Processing Unit (CPU), they act as bridges between the microchip and the base plate.

Microprocessor connectors
Join the microprocessor to the network of PGA connectors via microscopic wires.

CIRCUITS TO BE PRINTED

Small chips are manufactured, by "printing" circuits onto silicon wafers, using a technique called photoli-thography. The wafers are divided into "dices" containing the circuits. The dices are then cut and encapsu-lated in the connection and protection structures.

① The integrated circuit is designed.

② Using the photolithography process, the design is copied onto a silicon wafer.

③ The circuit is transferred onto a wafer. In the same wafer are identical circuits.

④ The integrated circuits are trimmed to separate them.

⑤ The circuit terminals are onto the sepa-rated circuits.

⑥ The protective enclo-sure is mounted into the circuits.

The computer

Although it was conceived as a laboratory instrument for carrying out complex calculations, in the 21st century the computer has become a fundamental part of people's lives. Computers have apparently infinite applications, covering everything from industrial processes to services, communication to entertainment. With the emergence of the Internet and email in the 1990s, computers became even more important to society. These two pillars of the globalized world could not exist without computers.

Inside a PC

A PC unit houses a labyrinth of cables, chips and circuits which are indecipherable to the average person. However, each part is clearly unique and carries out a specific function, albeit in connection with the other parts.

→

Power suply
Receives energy from an external sources and supplies it to the computer. The power supply has a dedicated fan to prevent overheating.

Video card
This electronic device enables certain information that is managed and processed by the computer to be displayed on a video device, such as a monitor. It also houses the connection port for this monitor.

29.76 tons
(27 metric tons)

The weight of ENIAC, considered to be the first computer. It was able to solve 5,000 sums and 360 multiplications in 1 second.

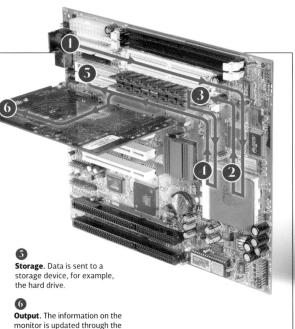
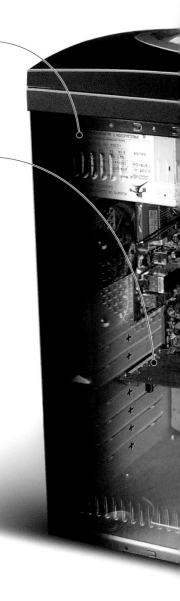

HOW A COMPUTER WORKS

A basic action routes information throughout the computer's components.

❶ Input. Data enters the computer through a keyboard, mouse, or modem and is interpreted by the appropriate circuit.

❷ Microprocessor. Controls all computer functions. It processes the entered data and carries out the necessary arithmetic and logic calculations.

❸ RAM (Random Access Memory). Temporarily stores all the information and programs used by the microprocessor.

❹ Processing. Data can travel back and forth from the CPU to the RAM several times until processing is complete.

❺ Storage. Data is sent to a storage device, for example, the hard drive.

❻ Output. The information on the monitor is updated through the video card.

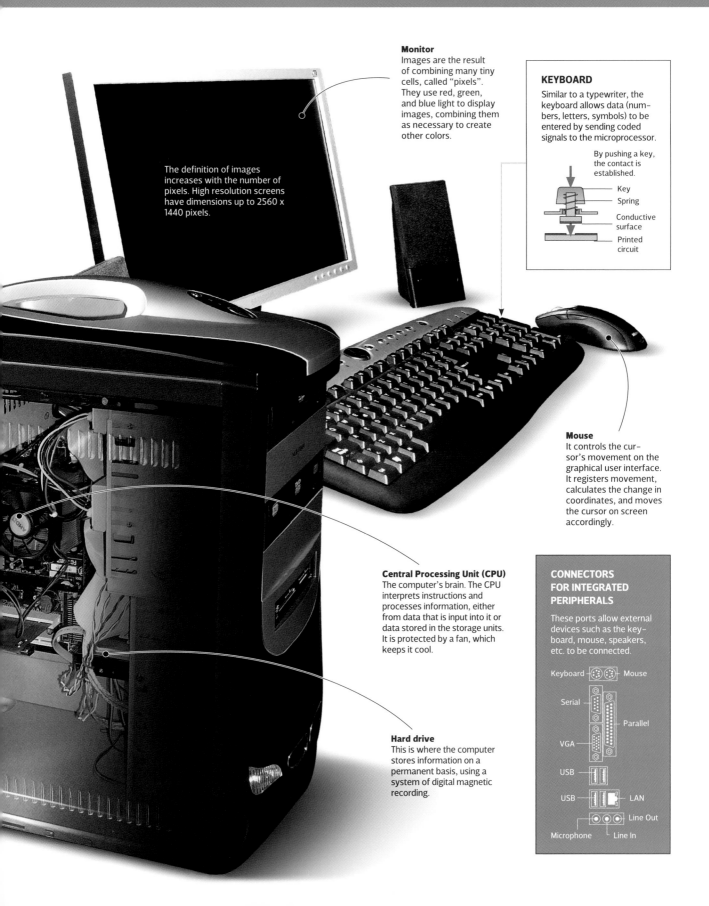

Monitor
Images are the result of combining many tiny cells, called "pixels". They use red, green, and blue light to display images, combining them as necessary to create other colors.

The definition of images increases with the number of pixels. High resolution screens have dimensions up to 2560 x 1440 pixels.

KEYBOARD
Similar to a typewriter, the keyboard allows data (numbers, letters, symbols) to be entered by sending coded signals to the microprocessor.

By pushing a key, the contact is established.

- Key
- Spring
- Conductive surface
- Printed circuit

Mouse
It controls the cursor's movement on the graphical user interface. It registers movement, calculates the change in coordinates, and moves the cursor on screen accordingly.

Central Processing Unit (CPU)
The computer's brain. The CPU interprets instructions and processes information, either from data that is input into it or data stored in the storage units. It is protected by a fan, which keeps it cool.

CONNECTORS FOR INTEGRATED PERIPHERALS

These ports allow external devices such as the keyboard, mouse, speakers, etc. to be connected.

Keyboard — Mouse
Serial —
— Parallel
VGA —
USB —
USB — LAN
Line Out
Microphone — Line In

Hard drive
This is where the computer stores information on a permanent basis, using a system of digital magnetic recording.

The digital camera

Photography, the technique of recording fixed images on light-sensitive surfaces, made huge advances with the development of the digital camera. Digital cameras are based on the principles of traditional photography, but while traditional cameras affix images onto film coated with chemical substances sensitive to light, digital cameras process the intensity of light and store the data in a digital format. Modern digital cameras generally have multiple functions and are able to record sound and video in addition to photographs.

Compact or DSLR

Digital cameras generally fall into these two categories. Compact cameras are user-friendly, making them easy to pick up and take pictures. DSLR (Digital Single-Lens Reflex) cameras are more complex and have a lens that can be changed.

→

IMAGE CAPTURE

In digital cameras the sensor is exposed to the light passing through the camera lens. They use a CCD (charge-coupled device) with a bayer filter or three separate image sensors (one each for the primary colors: red, green, and blue).

Object

Lens
Allows camera to focus on a consistent image.

Diaphragm
Regulates the amount of light entering the lens.

Shutter
Determines the exposure time.

CCD
Array of interconnected semiconductors.

Digital image
Appears upside down.

THE SENSOR THAT REPLACES FILM

The CCD (charge-coupled device) is a group of small diodes sensitive to light (photosites), which convert photons (light) into electrons (electric charges).

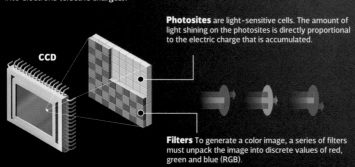

Photosites are light-sensitive cells. The amount of light shining on the photosites is directly proportional to the electric charge that is accumulated.

CCD

Filters To generate a color image, a series of filters must unpack the image into discrete values of red, green and blue (RGB).

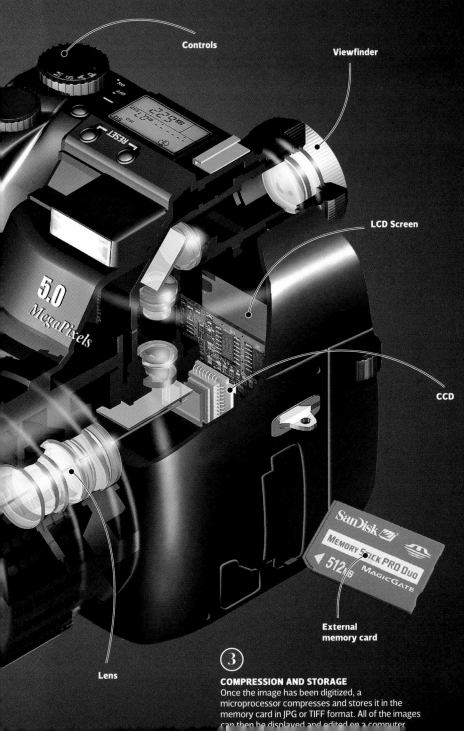

1972

was the year in which Texas Instruments first designed a filmless camera.

Controls

Viewfinder

LCD Screen

5.0 MegaPixels

CCD

External memory card

Lens

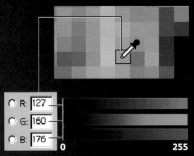

②

BINARY SYSTEM PROCESSING

To convert electrical charges from the photosite (analog) to digital signals, the camera uses a converter that codes them as pixels (colored dots)

ADDITIVE MIXTURE

Each pixel is colored by mixing values of red, green or blue (RGB). Varying quantities of each of these colors can reproduce almost any color of the visible spectrum.

R: 127
G: 160
B: 176

0   255

The value of each color can vary from 0 (darkest, closest to black) to 255 (the greatest color intensity).

RESOLUTION

is measured in PPI, or pixels per square inch— the number of pixels that can be captured by a digital camera. This figure indicates the size and quality of the image.

③

COMPRESSION AND STORAGE

Once the image has been digitized, a microprocessor compresses and stores it in the memory card in JPG or TIFF format. All of the images can then be displayed and edited on a computer.

↓ THE INVENTOR OF THE DIGITAL CAMERA In 1975 Eastman Kodak asked the American engineer Steven Sasson if he could build a camera that used solid state electronics and an electronic sensor to gather optical information. In 1978 Sasson was issued a patent for the first digital camera.

Virtual laser keyboard

Given that a simple wireless keyboard still baffles some people, a virtual laser keyboard might seem like a fantasy invention from a science fiction movie. Yet, virtual laser keyboards are already a reality and can be bought at an affordable price. With this device, users type on a virtual keyboard that can be projected onto a wide variety of surfaces. Far from being a technological fad or a gimmick with no real use, the virtual laser keyboard may be the answer to a serious problem: typing on mobile phones is often difficult because of the small size of keys.

15 minutes

The practice time needed to get the hang of the virtual keyboard, according to the manufacturer.

Writing in light

A tiny device, even smaller than a mobile phone, is all that is needed to generate the virtual keyboard, which can be projected onto any opaque surface. ↘

PDA ←→ Virtual keyboard

Operates up to a distance of 30 ft (9 m).

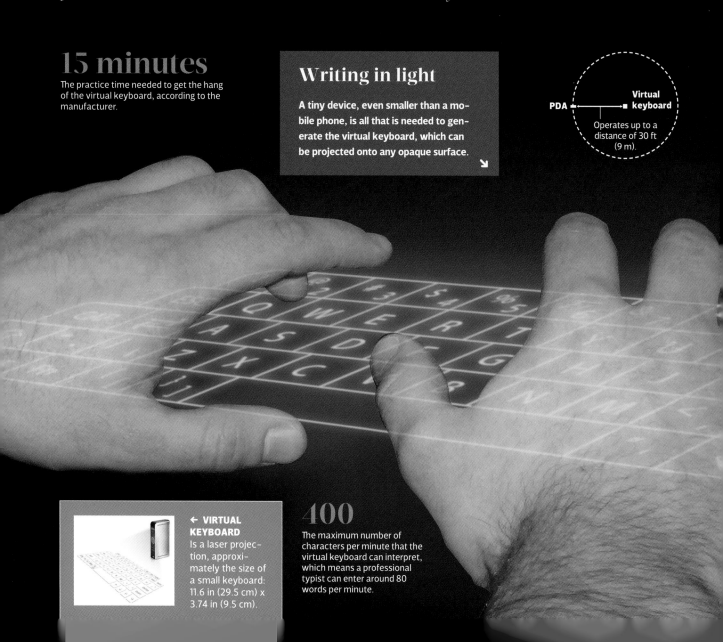

← VIRTUAL KEYBOARD

Is a laser projection, approximately the size of a small keyboard: 11.6 in (29.5 cm) x 3.74 in (9.5 cm).

400

The maximum number of characters per minute that the virtual keyboard can interpret, which means a professional typist can enter around 80 words per minute.

Projector
This is the heart of the virtual keyboard. It measures just 3.6 in (9.2 cm) x 1.4 in (3.5 cm) and weighs 0.2 lbs (90 g).

Projection window

Projection surface

HOW DOES IT WORK?

Although the user types on the virtual laser keyboard, detection is actually performed by an invisible infrared layer, located directly above the virtual keyboard.

The laser projector generates the virtual keyboard onto an opaque surface. At the same time, a diode generates an infrared layer, parallel to the keyboard, located just a few hundredths of an inch above the projection.

When the user presses one of the projected keys, the infrared light field is broken, producing an ultraviolet reflection, which is also invisible.

The reflection is picked up by a camera, which sends the signal to a chip. This chip calculates the position of the key that has been "pressed," based on the distance and angle of reflection.

The information is transmitted via an infrared Bluetooth connection to the PDA, which displays the selected character on its screen.

Virtual keyboard
Infrared layer

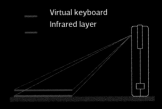

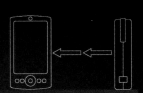

OTHER ALTERNATIVE KEYBOARDS

Ergonomic keyboards
There are numerous, strangely shaped models. However, they all promise to make typing more comfortable and, in many cases, less painful.

Ergonomic keyboards
Without a doubt, this is one of the most unusual keyboards on the market. In fact, it has no keys at all, just two domes that, using wrist movements, allowing the user to type 128 characters and use three mouse combinations.

DataHand ergonomic keyboard
This device fits perfectly into the palm of the hand and even has a built-in mouse. The idea is to reduce the stress caused to fingers by long days of typing. The user selects the keys that appear on the key assignment display.

Roll-up keyboard
This is similar to a standard keyboard in every way except one: it is flexible and can be rolled up, making it easy to travel.

Memory Stick

Although it was only widely adopted in 1998, the USB (Universal Serial Bus) flash drive has become the medium of choice for the transportation and temporary storage of data. USB memory sticks use flash memory to save information without the need for batteries. Flash drives are small, durable, lightweight, quick, reliable, and practical. Memory sticks are also versatile, for example, allowing a computer to be started using specific stored parameters.

10 years

In theory, this is the useful life of a USB flash drive device. Taking into account the speed of technological advances, it is likely that the majority of these devices will be obsolete before that time, and will no longer be in use.

Protecting cover
It protects the USB con-nection. Some flash drives have it integrated into the mechanism, protecting it without the danger of losing anything.

U3

A new technology that, aside from storing data in a flash memory device, allows users to carry installed programs. This new technology makes it possible for a user to temporarily personalize any computer.

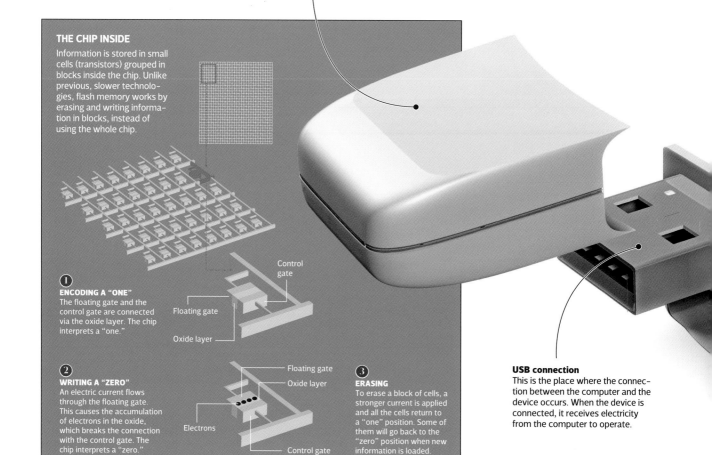

THE CHIP INSIDE

Information is stored in small cells (transistors) grouped in blocks inside the chip. Unlike previous, slower technolo-gies, flash memory works by erasing and writing informa-tion in blocks, instead of using the whole chip.

①
ENCODING A "ONE"
The floating gate and the control gate are connected via the oxide layer. The chip interprets a "one."

Control gate

Floating gate

Oxide layer

②
WRITING A "ZERO"
An electric current flows through the floating gate. This causes the accumulation of electrons in the oxide, which breaks the connection with the control gate. The chip interprets a "zero."

Floating gate

Oxide layer

Electrons

Control gate

③
ERASING
To erase a block of cells, a stronger current is applied and all the cells return to a "one" position. Some of them will go back to the "zero" position when new information is loaded.

USB connection
This is the place where the connec-tion between the computer and the device occurs. When the device is connected, it receives electricity from the computer to operate.

8 Mb

The capacity of one of the first external storage devices, released by IBM in 2000.

OTHER TECHNOLOGIES

Flash memory prompted the development of a number of other devices that perform the same function: the external storage of data and the transportation of data between different systems.

SD cards

These cards are used in a whole host of applications. Like USB drives, they work with flash memory, but are designed so that they can fit into ultra-compact or very small devices such as digital cameras, consoles and digital music players.

Cover

It is like a kind of shell that protects the device against impacts and accidents. Covers are even effective underwater.

Writing protection switch

It is a safety device similar to floppy discs. It allows information to be read, but not written or erased.

LED

A small light that turns on when the device is operating, either while reading or writing files.

Controller / Driver

The brain of the memory stick, the "great controller." It does the actual reading and writing of files and saves the memory in case the electric flow is interrupted.

Memory chip

It is the place where the information is stored, where it can be accessed by a computer. It is located on the back of the memory stick.

Flash memory: the key

Flash memory has several characteristics that have revolutionized the storage capacity of small devices such as cameras and mobile phones: it is non-volatile, which means it has no moveable components (making it more robust), the memory is not erased when the electric supply is cut, and it enables multiple operations, making it faster.

E–paper

A few years ago, the idea of electronic screens as thin as a sheet of paper and flexible enough that they could be rolled up and folded might have seemed incredible. However, this technology can now be found in some ebook readers and mobile phones. It is also revolutionizing old technology, adding new functionality to watches. Additional benefits of e–paper screens include extremely low energy consumption and excellent visibility from any angle and in any environment, even in direct sunlight.

Paper–thin

The main advantage of e-paper screens is their thickness and flexibility. Screens 0.05 in (1.2 mm) thick are already on the market.

0.012 in (0.3 mm)

The thickness of a prototype electronic paper screen presented by E Ink Corporation. That is about half the thickness of a credit card.

↓ **CHALLENGES** Black and white books print beautifully on e-paper. Developing an efficient color screen (prototypes have already been made) and improving the refresh rate so that it is possible to display videos smoothly are the challenges facing this emerging technology now.

2.19 in (5.56 cm)

4 in (10 cm)

Headphone socket

USB port
Enables connection to a PC, modem, printer, or any other hardware.

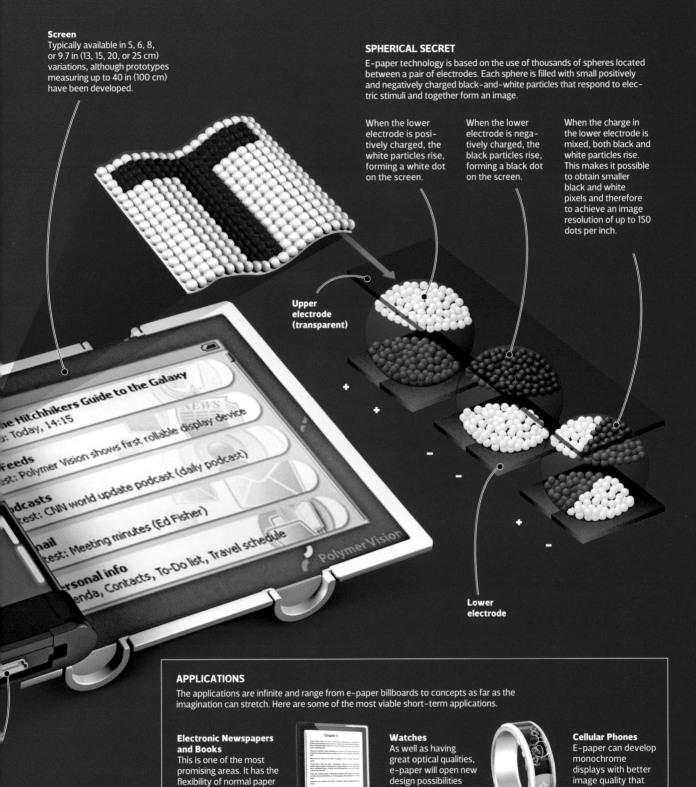

Screen
Typically available in 5, 6, 8, or 9.7 in (13, 15, 20, or 25 cm) variations, although prototypes measuring up to 40 in (100 cm) have been developed.

SPHERICAL SECRET

E-paper technology is based on the use of thousands of spheres located between a pair of electrodes. Each sphere is filled with small positively and negatively charged black-and-white particles that respond to electric stimuli and together form an image.

When the lower electrode is positively charged, the white particles rise, forming a white dot on the screen.

When the lower electrode is negatively charged, the black particles rise, forming a black dot on the screen.

When the charge in the lower electrode is mixed, both black and white particles rise. This makes it possible to obtain smaller black and white pixels and therefore to achieve an image resolution of up to 150 dots per inch.

Upper electrode (transparent)

Lower electrode

The Hitchhikers Guide to the Galaxy
u: Today, 14:15

Feeds
est: Polymer Vision shows first rollable display device

odcasts
est: CNN world update podcast (daily podcast)

nail
est: Meeting minutes (Ed Fisher)

rsonal info
nda, Contacts, To-Do list, Travel schedule

Polymer Vision

APPLICATIONS

The applications are infinite and range from e-paper billboards to concepts as far as the imagination can stretch. Here are some of the most viable short-term applications.

Electronic Newspapers and Books
This is one of the most promising areas. It has the flexibility of normal paper plus all the applications of an electronic screen.

Watches
As well as having great optical qualities, e-paper will open new design possibilities because of the screen's flexibility.

Cellular Phones
E-paper can develop monochrome displays with better image quality that have great visibility from any angle or in any environment.

3-D Printers

The recent rise of 3-D printers brought an inexpensive and practical alternative to large industrial modeling machines. Of similar size to—or even smaller than—a photocopier, they can quickly and easily create three-dimensional objects. Models can be printed in a wide range of formats, from the very simple to the highly complex, and can even be printed in color. They are controlled by a normal computer using special 3-D modeling software and are very efficient because they have the option to reuse waste materials.

The printer

Can build 3-D objects from 8 to 12 in (20–30 cm) in length, depending on the model, using a special powder of fine particles and an adhesive material that acts like glue. ↘

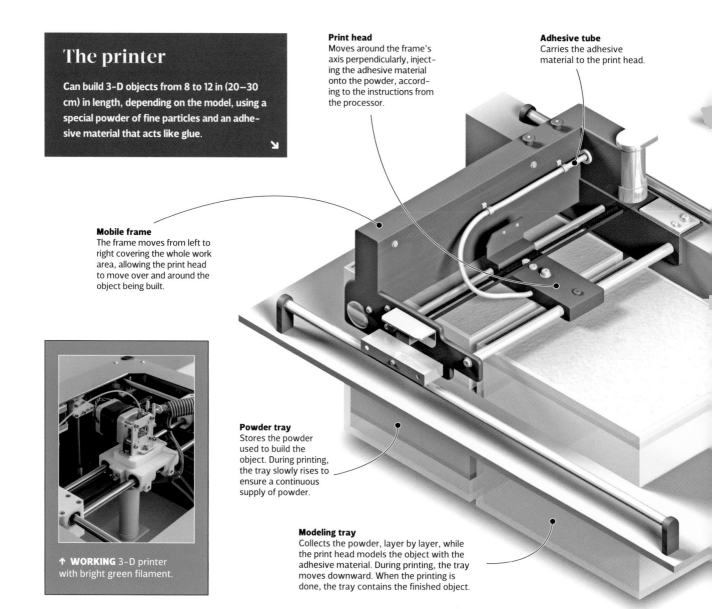

Print head
Moves around the frame's axis perpendicularly, injecting the adhesive material onto the powder, according to the instructions from the processor.

Adhesive tube
Carries the adhesive material to the print head.

Mobile frame
The frame moves from left to right covering the whole work area, allowing the print head to move over and around the object being built.

Powder tray
Stores the powder used to build the object. During printing, the tray slowly rises to ensure a continuous supply of powder.

Modeling tray
Collects the powder, layer by layer, while the print head models the object with the adhesive material. During printing, the tray moves downward. When the printing is done, the tray contains the finished object.

↑ **WORKING** 3-D printer with bright green filament.

COMPUTERIZED SCULPTING

3-D printing builds objects layer by layer, from the base to the top. It is a slow process, but it is quicker and cheaper than building models in the traditional way.

❶ THE DESIGN
Created on a computer screen, using a 3-D modeling program.

❷ THE BASE
The print head sprays a fine layer of powder into the modeling tray.

Powder tray

Powder

Modeling tray

Powder tray

❸ THE PRINT PROCESS
The quick-drying adhesive material is then injected onto the layer of powder. This process is repeated for each layer of the object.

Modeling tray

Powder tray

❹ FINISHING STAGES
Once completed, the object is removed from the modeling tray. Finally, it is dipped in various liquids to achieve the desired rigidity.

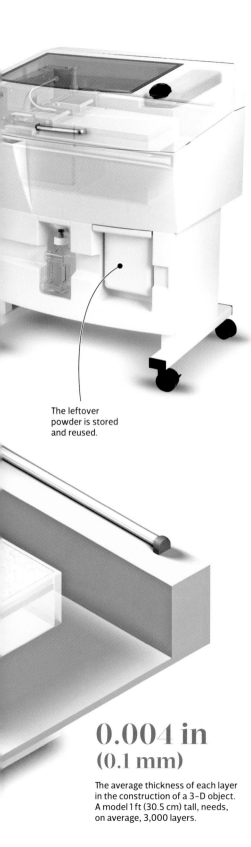

The leftover powder is stored and reused.

0.004 in (0.1 mm)

The average thickness of each layer in the construction of a 3-D object. A model 1 ft (30.5 cm) tall, needs, on average, 3,000 layers.

Barcode

It is unlikely that the globalized mass market could operate at its current level without the barcode. A barcode is a two-colored label that encodes certain information about a given product . When the barcode is read by an optical scanner, the product can be identified within a fraction of a second. People frequently encounter barcodes in daily life, especially when shopping in a supermarket or other store, but barcodes have a wide range of other applications, including logistics, transportation and the distribution of goods.

Speed reading

Barcodes can be read by an optical reader or scanner which is able to decipher the information contained in the barcode label in less than a second.

↘

→ **UTILITIES**
The barcode is very visible and can be seen on a wide range of products, from supermarket goods to this book. In addition to identifying the product, this barcode is used to control goods, inventories, production, quality and more.

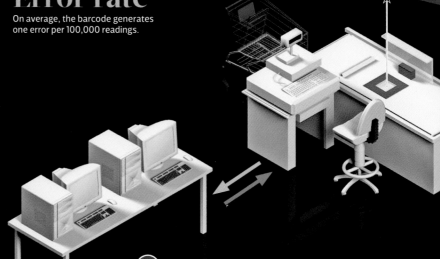

② **Laser**

Error rate

On average, the barcode generates one error per 100,000 readings.

④ **Database**

2D code

Matrix code

↑ **OTHER CODES** EAN-13 is the most widely used barcode in the world. Other codes (even in 2D) make it possible to include specific information for different activities.

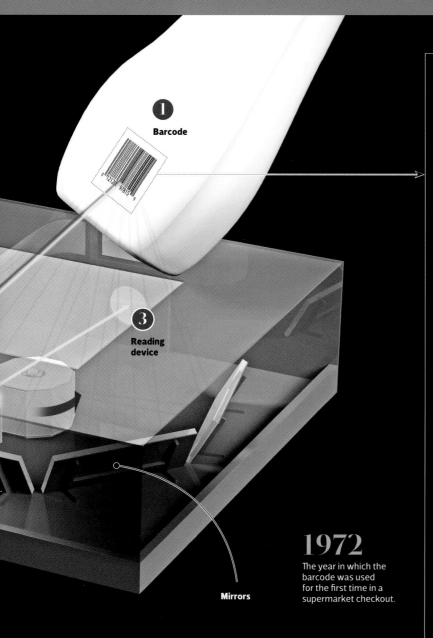

1

Barcode

3

Reading device

Mirrors

1972

The year in which the barcode was used for the first time in a supermarket checkout.

THE CODE

The black bars and white spaces are a codification of the 13 digit number accompanying all the barcodes associated with products. These numbers provide three types of information:

Laser
The red laser emitted by the reader scans the code.

Reflection
The black bars absorb the red light, while the white bars reflect it.

0 121200 97815 9

The first three digits detect the country of origin.

The six following digits give information about the manufacturer.

The other three digits detect the article.

The last digit is a check digit.

Symbology
Each of the 13 digits that form the number is encoded using the binary system, with ones and zeros. The ones and zeros are defined by the width of the bars and spaces.

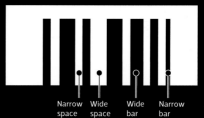

Narrow space

Wide space

Wide bar

Narrow bar

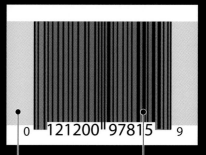

0 121200 97815 9

Quiet zone
This is an empty space that enables the reader to distinguish the code from the rest of the product label.

Coding zone
Contains information about origin, the company and the product itself.

 1

The product is swiped in a vertical position in front of a reader that illuminates the barcode with a laser beam.

 2

The red laser emitted by the reader scans the code. The black bars absorb the red light, while the white bars reflect it.

 3

The reflections are picked up by the reading device which sends a signal to a decoder. The decoder converts the bars into a binary numeric code, and then into a decimal code.

 4

The processor compares the numeric code read with those in its database, and thus identifies the product. When it identifies the product, the processor obtains other information, which is not included in the barcode, such as its price and name.

Fiber Optics

Today's most efficient and widespread system for the transportation of information is based on a few simple optical principles. Researchers discovered that optical fiber is not only lighter and more economical and versatile than traditional copper wire, but also allows for a much greater and faster flow of data. That data travels through filaments the thickness of a human hair, converted into pulses of light. Fiber optics also have great implications for the medical field, where it has allowed formerly invasive operations and examinations to be carried out with a minimum of suffering for patients.

An "illuminated" cable

A phenomenon known as "total reflection" lets light travel through a fine tube of glass or plastic (fibers), traveling long distances with minimal loss. Fibers are also grouped into numerous sets to form wire. ↘

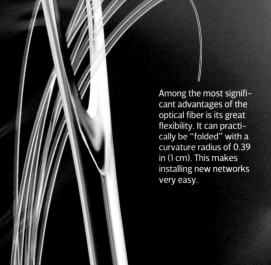

Among the most significant advantages of the optical fiber is its great flexibility. It can practically be "folded" with a curvature radius of 0.39 in (1 cm). This makes installing new networks very easy.

↑ **APPLICATIONS** Telecommunications is undoubtedly the area in which the use of optical fibers is most widespread. However, they are used in a number of other fields:

Medicine
Ultra-fine instruments can be made from strands of optical fiber and lenses, making it possible to examine objects through a small hole. These are known as endoscopes, and they help make diagnostic observations and surgical operations.

Industry
Endoscopes also have numerous industrial applications. Their ability to bend makes it possible to guide a beam of light to a point outside the line of vision.

THE OPTICAL FIBER

It consists of a core of glass or plastic coated with a layer of a similar material, but of lower refractive index.

Rubber cover

Coating

Nucleus

Inside the optical fiber the light bounces against the walls of the nucleus reflecting the entire way. In this way, it can carry information across large distances without suffering any losses.

The approximate thickness of an optical fiber is 0.004 in (0.1 mm).

24,233 mi
(39,000 km)

The length of SEA-ME-WE 3, the longest fiber optic cable in the world, which connects the Far East and Southeast Asia with the Middle East, Africa and Europe.

THE TRIP OF THE LIGHT

The transmission of data through fiber optics begins with an electrical signal that is converted into light, and then, at the end of the trip, converted back into an electric signal.

① A computer, a telephone, a radio transmitter or a television generates a binary or analog electrical signal.

② An encoder interprets the signal and converts it into light signals with the help of a diode (an LED or a laser).

③ The pulses of light generated by the diode travel through optical fiber cable.

④ Since the signal fades with distance —without degrading its characteristics—, optical regenerators, located at specific distances in the wire, amplify it.

⑤ A photodiode capture signal lighting at the e cable and through a c transforms it into elec analog or digital signa reaches the receiving computer, phone, etc

Smart clothing

In the coming years, our clothing will undergo some of the most dramatic and surprising evolutions since humanity first began wearing clothes. Smart fabric and computerized apparel already exist, and some examples have already entered the market and are available for mass consumption. Among them, are materials that integrate features that would have been hard to imagine just a few years ago—for example, clothing that not only informs the wearer of the body's response to physical activity, but also modifies itself to improve performance.

Information in real time

Clothes constructed from fabrics with integrated minisensors and imperceptible electrical circuits can determine the wearer's heart rate, levels of oxygen and other gases in blood, calories consumed, and breathing rate. →

5,000,000

The number of calculations per second performed by the Adidas-1 microchip.

↑ **DIVERSE USERS** Smart apparel is obviously of great benefit to athletes, but it is also useful to patients with chronic illnesses who need to frequently monitor their condition, including those with diabetes and heart problems.

SMART FABRICS

A product of new developments in nanotechnology, smart fabric shows surprising features that will be widely used in the next few years.

Colorful
A special fiber made of plastic and glass can be used with electronic circuitry to modify the way the fabric reflects light, thereby changing its color.

Comfortable
Fabrics that eliminate sweat, keep the skin dry and eliminate odors already exist. Similarly, there are materials that can provide ventilation or warmth in accordance with the outside temperature.

Resistant
Fabrics that do not get wrinkled, are resistant to stain, and keep their shape after many years of wear and washing have also been developed.

Antistatic
Fabrics that remove static electricity. They prevent the buildup of hair, pollen, dust, and other potentially harmful particles for people with allergies.

Antimicrobial
Fabrics that block the growth of viruses, fungi, bacteria, and germs.

Microphone

Fiber-optic cable

Sensors

Sensors

Database

Transmitter

When a person is running, the body absorbs three to four times the person's weight each time a step is taken. Smart shoes help absorb this enormous force and protect the most vulnerable areas. They also provide comfort and stability.

Chlorine

Is an element found in the fibers of fabrics that repel germs. One of its properties is that it destroys bacterial cell walls. It is also the basis of bleach, which is frequently used in disinfectants.

PERFECT STEPS

The Adidas-1 shoe, a project three years in the making, can determine the athlete's weight, stride, and surrounding terrain to adjust the shoe's tension accordingly.

①

Inside the hollow heel, the components of the shoe generate a magnetic field.

Magnetic field

②

While running, the foot hits the heel of the shoe and modifies the magnetic field.

③

A sensor that can perform up to 1,000 readings per second detects each modification and sends that information to the microchip.

④

The microchip determines the appropriate ten-sion for the heel and sends the information to the motor.

⑤

The motor, rotating at 6,000 rpm, moves the screw, which in turn strengthens or relaxes the heel. The entire process is repeated with each step.

Soft heel

Firm heel

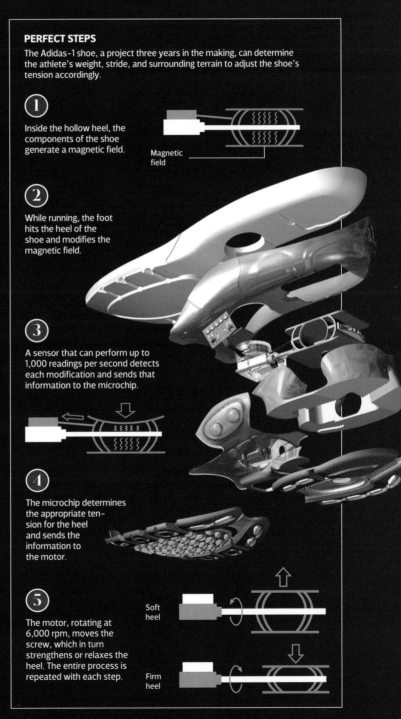

ARCHITECTURE

Christ the Redeemer

The majestic statue of Christ the Redeemer stands in the center of Rio de Janeiro, Brazil, on the peak of Mount Corcovado. The statue, with its open arms, stands on a pedestal of just 50 ft (15 m). Located 2,330 ft (709 m) above sea level, it is surrounded by the spectacular Tijuca Forest, the world's largest urban wood. Inaugurated in 1931, this 125 ft (38 m) tall statue was built to commemorate a hundred years of Brazilian independence. Today it is one of the main tourist attractions in Rio de Janeiro, and indeed the country.

CREATORS OF THE CHRIST

The project's creator was engineer Heitor da Silva Costa; the artist Carlos Oswald was responsible for the final design; and the French sculptor Paul Landowski made the head and hands.

← **The sculptor**
Landowski used the hands of poet Margarita Lopes de Almeida as a model.

Construction

The statue took five years to build and was inaugurated on 12 October 1931. Beside the site's exposure to strong winds, the scaffolding barely fit on the pedestal and the open arms and bowed head required a feat of engineering.

↘

LED's
The access path has 300 lights, operated by remote control, that highlight the statue.

The lookout
Christ the Redeemer receives 1.8 million visitors a year. Visitors enjoy one of the most magnificent panoramic views of the city of Rio and Sugarloaf Mountain.

↑ **RIO DE JANEIRO** Aerial view from a helicopter of Rio de Janeiro with Mount Corcovado and the statue of Christ the Redeemer. Sugarloaf mountain is visible in the background.

The arms
The people of Rio voted for the Christ to have his arms open, in a welcoming gesture. Each arm measures 89 ft (27 m) and weighs 30 tons.

Head
Slightly bowed, the head is made up of 50 pieces. Christ is looking east toward Guanabara Bay, and further to Jerusalem.

The outside
It is covered by mosaics made from soapstone from Minas Gerais.

The heart
Situated on the 11th floor, on the outside of the statue, the heart is 4.27 ft (1.3 m), which is proportionally rather small, compared with the human body.

The interior
A skeleton of beams and stairways fills the interior.

The structure
It is made of reinforced concrete and has 12 floors inside.

Art deco

Rio's magnificent statue is the world's biggest Art Deco monument and a Brazilian icon.

The pedestal
This is 26 ft (8 m) high and has a chapel dedicated to Our Lady of Aparecida inside. The chapel holds 23 people and hosts weddings and baptisms.

Access
Escalators and three panoramic elevators were installed in 2002. Access on foot is by a stairway with 220 steps.

Notre Dame

Starting in the 10th century, Europe underwent a period of relative economic growth. The large cities expanded and socio–political structures went through profound reforms. In this context, cities began to be build great Christian cathedrals. A gothic emblem and an icon of Paris, Notre Dame's cathedral has a great number of treasures, both inside and outside. It was constructed between the 12th and 15th centuries in the same spot where Pagan and Christian temples had been before. At that time, it was the largest cathedral in Christendom.

A new style

The Gothic style, despite being messy and undefined in many aspects, has a set of unique features. For example, it shows a deep concern for the entrance of light into the building, so they used large decorated windows. It is also possible to identify three characteristic elements: ogive, ribbed vault and flying buttress.

Roof
Built in the 14th century and weighs 231 tons.

Flying buttress
Characteristic of Gothic architecture, they are external structural elements with the shape of half arches that support the weight of the dome and transmit it to a buttress mounted to the wall of the lateral nave.

The apse
Is the oldest section in Notre Dame dating back to the 12th century.

↓ **THE CHIMERAS** One of the characteristic features of the cathedral of Notre Dame is the fantastic birds, demons and monsters that decorate it. This bold artwork dates to the great restoration in the 19th century and did not exist in the Middle Ages.

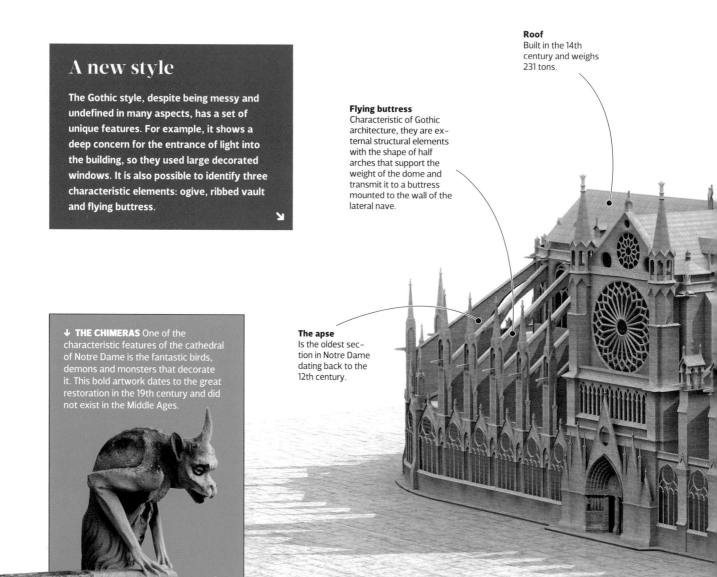

FLOOR

It is simple, Roman cross style with a large central nave and a choir crossed by a large transept and two lateral naves.

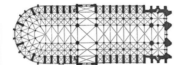

West facade

THREE DISTINCTIVE ELEMENTS

This Gothic cathedral has at least three characteristic elements:

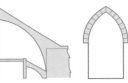

Flying buttress   Ogive   Ribbed vault

Pinnacle
Built during the restoration of the cathedral in the 19th century.

The towers
They reach a height of 226 ft (69 m). The North Tower was finished around 1240 and the South Tower a decade later.

West facade
The most famous facade of the cathedral. It was built between 1200 and 1250.

Chimera gallery
It links the two towers of the facade by means of a colonnade decorated with chimeras.

Rose window
Measuring 31.5 ft (9.6 m), it is the smallest of the three decorating the cathedral and crowns Virgin Mary's balcony.

Kings' gallery
28 statues that are 9.8 ft (3 m) tall representing 28 generations of Judah's kings that preceded Christ.

Saint Anne's portal
Devoted to Virgin Mary's mother, it is a remnant of the church that first occupied the spot where the Cathedral was built. It dates back to the 12th century.

Final Judgement portal
It has scenes of Christ's martyrdom and the Final Judgement.

The Virgin portal
It is devoted to the life of the Virgin, the main figure venerated in the temple.

Saint Peter's Basilica

When Christianity became the official religion of the Roman Empire in the 4th century, the basilica shape was used to build the most important churches. Saint Peter's Basilica, standing in Vatican City, is the preferred site for the Pope to celebrate religious rites. It was constructed in the 17th century, when it was shaped like a Latin cross. It was built on the site of a former basilica that stood on Saint Peter's tomb. Saint Peter was the first Pope, who became a martyr in the 1st century.

The big room

If there is something outstanding about the Basilica, it is the simplicity of it shapes: a wide rectangular room with one or more naves surrounded by Roman pillars, with an apse at the opposite side of the entrance. This apse is a semi-circular shaped section.

Apse

Normally, during a service, religious authorities are located in this space, while the devotees stay on the naves.

Dome
The largest dome in the world at 448 ft (136.57 m) high. It is designed by Michelangelo and completed after his death.

↓ **THE PLAN**

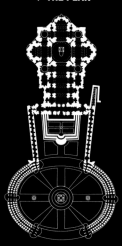

↑ **THE PIETÀ** The sculptural and religious treasures that the Saint Peter's Basilica holds are endless. One of them, *The Pietà* of Michelangelo, carved in marble, is one of the most famous and is a must-see for those visiting the church.

Pope's altar
It is located under the dome, protected by the huge baldaquin of Saint Peter, 98.5 ft (30 m) tall, that recalls Solomon's Temple. It is located above Saint Peter's tomb.

The naves
There are three naves. The central nave is 614 ft (187 m) by 148 ft (45 m), the Gospel nave. The naves are separated by huge pillars.

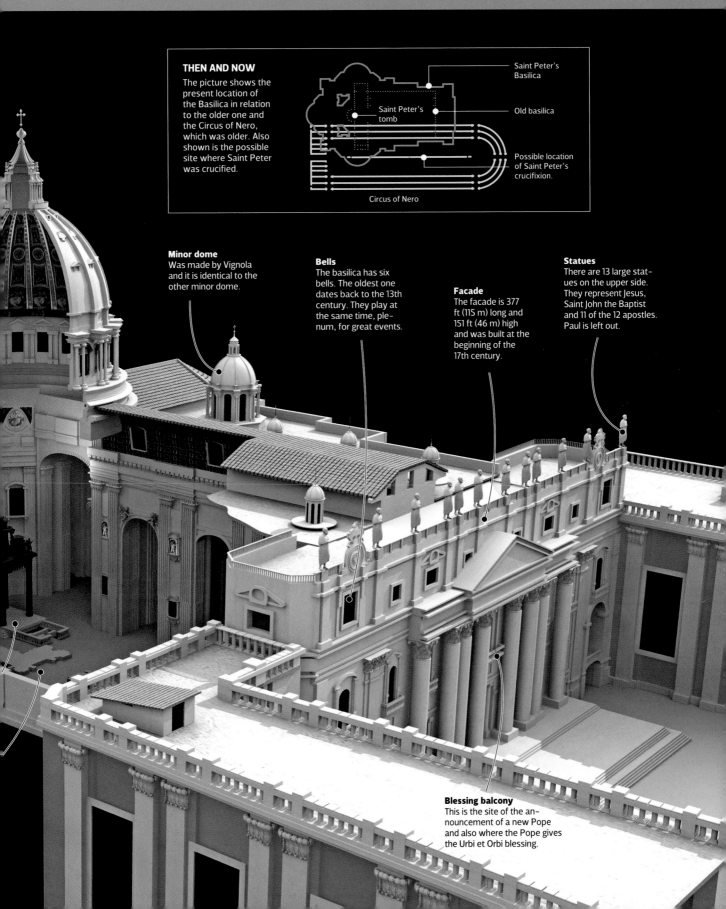

THEN AND NOW
The picture shows the present location of the Basilica in relation to the older one and the Circus of Nero, which was older. Also shown is the possible site where Saint Peter was crucified.

Saint Peter's Basilica

Saint Peter's tomb

Old basilica

Possible location of Saint Peter's crucifixion.

Circus of Nero

Minor dome
Was made by Vignola and it is identical to the other minor dome.

Bells
The basilica has six bells. The oldest one dates back to the 13th century. They play at the same time, plenum, for great events.

Facade
The facade is 377 ft (115 m) long and 151 ft (46 m) high and was built at the beginning of the 17th century.

Statues
There are 13 large statues on the upper side. They represent Jesus, Saint John the Baptist and 11 of the 12 apostles. Paul is left out.

Blessing balcony
This is the site of the announcement of a new Pope and also where the Pope gives the Urbi et Orbi blessing.

The Forbidden City

The Forbidden City was the center of power of the Ming (1368–1644) and Qing (1644–1911) dynasties for more than 500 years. It is located in the center of old Beijing and is not exactly a city, but rather the largest palace complex in the world with an area of 447 mi² (720 km²) and was designed as the center of the concentric sectors the Ming dynasty had divided Beijing into. The population lived outside the city while highly qualified government employees were located inside. The center of Beijing, or the Forbidden City, was only for the Emperor. Entering this area was punished with death.

Imperial residence

This Imperial residential complex was built between 1406 and 1420. It housed 24 emperors of the Ming and Qing dynasties. It is divided into two main parts: the Outer Court, in the south side of the city and the Inner court, at the North. This was the place where the Emperor and his family lived together with their eunuchs and servants. ➘

1,000,000
Workers contributed to the construction of the city during the first phase. Later this work was completed by 100,000 artisans.

932 miles
(1,500 km)

The cut logs were transported over a distance of 932 miles (1,500 km) from the Sichuan woods before reaching Beijing.

DETAILS AND DECORATIONS

The majority of the buildings are made of wood. The roofs are supported by robust columns. Roofs and walls are decorated with amazing detail.

Sculptures
The sculpted animals had a symbolic value: bronze lions guard the palace; dragons bestow goodness and rightness; cranes bring longevity for the monarch and the empire.

 1 **THE THREE PAVILIONS**
The Harmony Halls were the most important buildings of the complex.

 2 **ROOMS**
More than nine thousand pavilions and wooden rooms spread symmetrically towards North and South.

 3 **DEFENCES**
The Forbidden City was surrounded by a ditch with water that was 20 ft (6 m) deep and 170 ft (52 m) wide.

 4 **ELITE RESIDENCES**
They housed the aristocracy. The roofs were made of yellow glazed tiles, the color of the Empire.

 5 **WALLS**
The city is surrounded by a ditch and a wall 33 ft (10 m) tall, containing a watchtower at each corner.

Tiles
All the rooms, without any exceptions, are decorated with yellow crystal tiles, the imperial color.

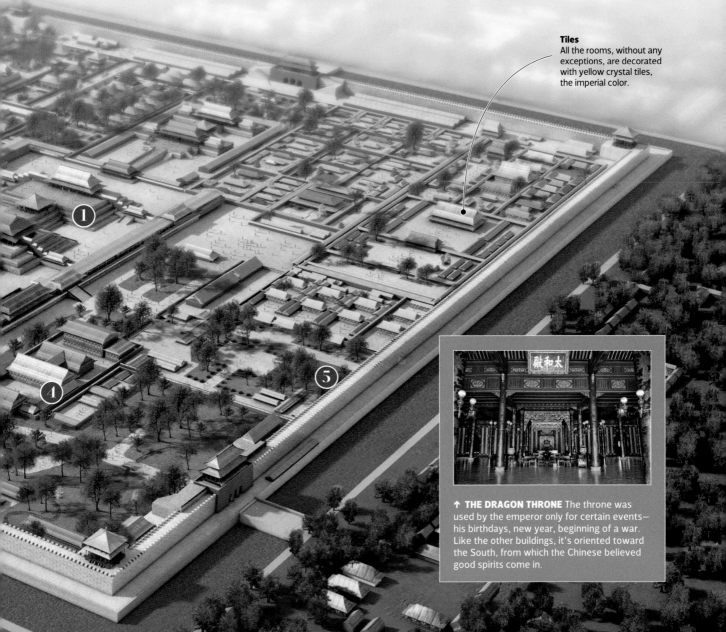

↑ **THE DRAGON THRONE** The throne was used by the emperor only for certain events—his birthdays, new year, beginning of a war. Like the other buildings, it's oriented toward the South, from which the Chinese believed good spirits come in.

The Reichstag

At the end of the 20th century the building housing the Reichstag, a historical edifice and one of the most visited places in Berlin, was fundamentally rebuilt in neo-Renaissance style to hold the Bundestag–the Federal German parliament. The original building, built at the end of the 19th century at the time of German unification, was left in ruins by the Second World War. It sat, mostly unused, for decades and was redesigned by British architect Norman Foster in the 1990s to celebrate German reunification.

↑ **INTERIOR** The inside was completely redesigned between 1994 and 1996, using concrete, glass and steel. Brightness, energy efficiency and modernity prevail, with a range of colors in the different areas.

The main facade
Designed by Paul Wallot in 1895 in neo-Renaissance style, the main facade's entrance is decorated with a staircase and Corinthian columns.

Foster's mark

Designed by Paul Wallot, the building was finished in 1894. However, its current appearance owes to the restoration undertaken by the British architect Norman Foster, who won a public contest held in 1993.

→

The plenary hall
Located in the center of the building, this hall is approximately 3,937 ft² (1,200 m²). Completely remodelled, while respecting its historic heritage, it is presided over by the Federal Eagle. Sessions can be observed by the public from the upper dome.

DEM DEUTSCHEN VOLKE

A TURBULENT HISTORY

The building's vicissitudes reflect Germany's turbulent history in the 20th century. Designed to house the Parliament of the German Empire, it lost its function with the rise of the Nazis and did not become Parliament's seat again until 1999.

The fire of 1933
In 1933, shortly after Hitler was appointed Chancellor, a fire in the building, whose origins were unclear, caused serious damages. Subsequently, it was no longer used for parliamentary sessions.

World War II
The bombings and fighting at the end of the war left the building in a sorry state. Although it was rebuilt in the sixties, it was not used as the seat of the Bundestag until after reunification.

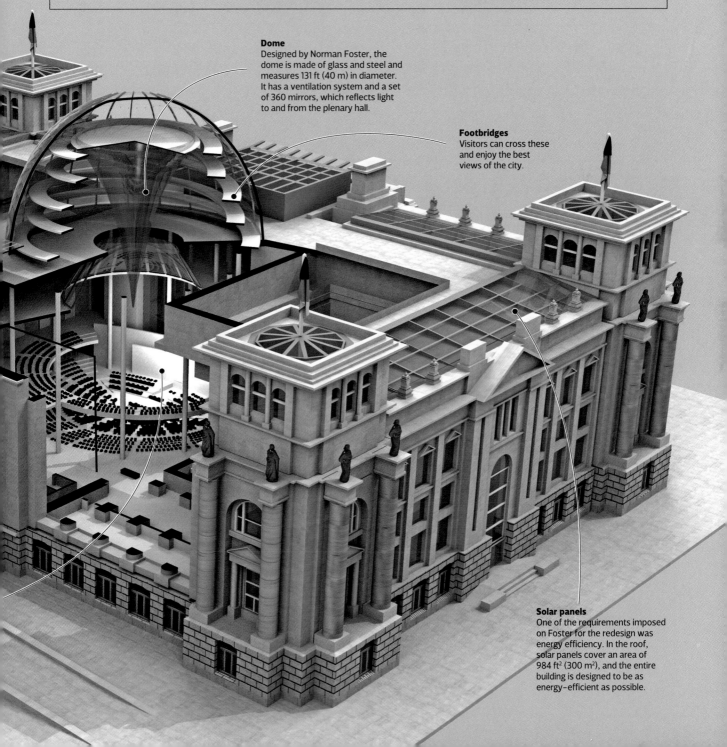

Dome
Designed by Norman Foster, the dome is made of glass and steel and measures 131 ft (40 m) in diameter. It has a ventilation system and a set of 360 mirrors, which reflects light to and from the plenary hall.

Footbridges
Visitors can cross these and enjoy the best views of the city.

Solar panels
One of the requirements imposed on Foster for the redesign was energy efficiency. In the roof, solar panels cover an area of 984 ft^2 (300 m^2), and the entire building is designed to be as energy-efficient as possible.

Hagia Sophia

The Byzantine Empire emerged after the capital of the Roman Empire moved from Rome to Constantinople (now Istanbul) in the 4th century, splitting the Roman Empire into Eastern and Western parts. Its most prosperous times took place around the 6th century, when Emperor Justinian reclaimed the ancient Roman Empire after taking over most of the Mediterranean. Justinian decided to build some of the most monumental examples of Byzantine architecture. They combined Roman and Middle Eastern influences with a deep sense of religion.

Saint Sophia

The Emperor Justinian ordered the building of Hagia Sophia, a beautiful Christian cathedral with a huge central dome that remained unsurpassed for the next thousand years. In 1453, Hagia Sophia became a mosque. In 1931, the religious building became a museum. ↘

THE GREAT SOLUTION

Building a dome was always a problem for builders in the old days. In Hagia Sophia, they designed a complex force transfer system as can be seen from the arrows.

Half-domes
Help support the central dome and, in turn, lie on other half-domes and vaults that transmit the thrust to the facades.

Exterior
Pyramidal appearance, the exterior was all mortar, except the towers, which were made of stone. Its sober decoration contrasts with the luxury of the interior.

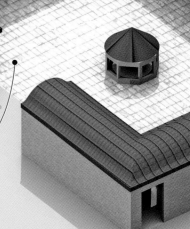

↑ **SHINING** Hagia Sophia is still radiant. A deep restoration, lasting two decades, was completed in 2010. The restoration restored the golden color to the mosaics covering its dome.

Atrium
It has five arcades with classical pillars and a great holy water receptacle.

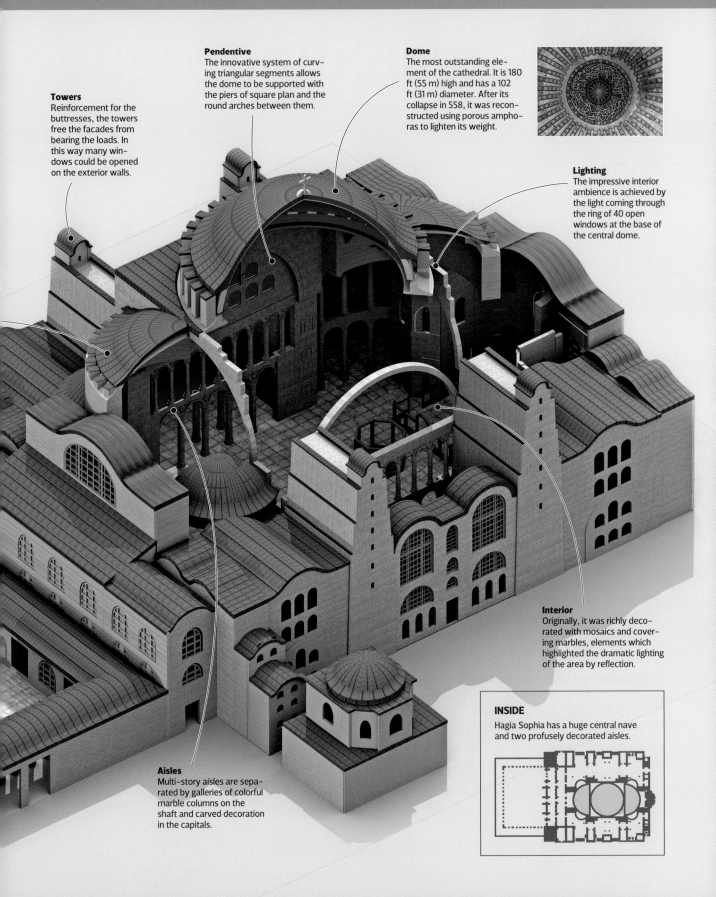

Towers
Reinforcement for the buttresses, the towers free the facades from bearing the loads. In this way many windows could be opened on the exterior walls.

Pendentive
The innovative system of curving triangular segments allows the dome to be supported with the piers of square plan and the round arches between them.

Dome
The most outstanding element of the cathedral. It is 180 ft (55 m) high and has a 102 ft (31 m) diameter. After its collapse in 558, it was reconstructed using porous amphoras to lighten its weight.

Lighting
The impressive interior ambience is achieved by the light coming through the ring of 40 open windows at the base of the central dome.

Interior
Originally, it was richly decorated with mosaics and covering marbles, elements which highlighted the dramatic lighting of the area by reflection.

INSIDE
Hagia Sophia has a huge central nave and two profusely decorated aisles.

Aisles
Multi-story aisles are separated by galleries of colorful marble columns on the shaft and carved decoration in the capitals.

Belz Great Synagogue

The origin of the synagogue as a place for meeting, praying and studying for the Jews, believed to be the oldest monotheistic religion, is not yet clear. Synagogues probably already existed in 700 B.C. Built in the 1980s by the Belz Hasidic community and consecrated in year 2000, Jerusalem's Belz Great Synagogue is one of the world's largest synagogues. It can seat a congregation of 6,000. It belongs to the Hasidic (Orthodox) branch of Judaism and has one of the most impressive arks.

Chandeliers

There are nine chandeliers. They are 18 ft (5.5 m) high with a diameter of 11 ft (3.4 m). Each one has 200,000 pieces of Czech crystal.

Free style

There is no pre-set architectural style for synagogues. Normally they have the prevailing style of the site and historical moment when they were built. However, there is a series of elements that must be respected and that usually evoke Solomon's Temple. →

Study

No synagogue lacks one or more spaces for studying.

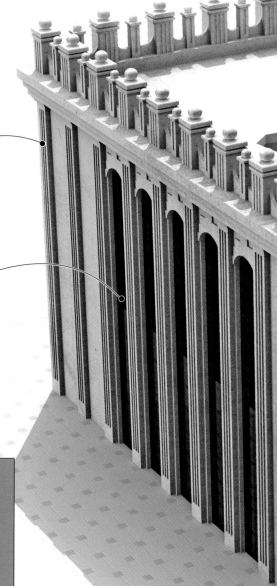

Main sanctuary
Only used for the most important celebrations and for Shabbat.

Entrances
The synagogue has four entrances accessible from the four streets surrounding the building in the hilltop neighborhood.

← **INSPIRATION** The Great Synagogue of Belz (Jerusalem) was inspired by another great Jewish temple built in 1843 in Belz (Ukraine). That synagogue, able to seat a congregation of 5,000, was destroyed during the Nazi invasion in 1939.

ESSENTIAL

These are the essential common elements present in all synagogues:

① MENORAH
The seven-branch candelabra, possibly inspired by the burning bush that Moses witnessed on Mount Sinai.

② TORAH SCROLLS
Sacred Jewish texts.

③ ARK
Contains the Torah scrolls.

④ RABBI'S CHAIR

⑤ BIMAH
The elevated platform from which the Torah is read.

⑥ PULPIT

⑦ SANCTUARY
The place where the main ceremonies are performed.

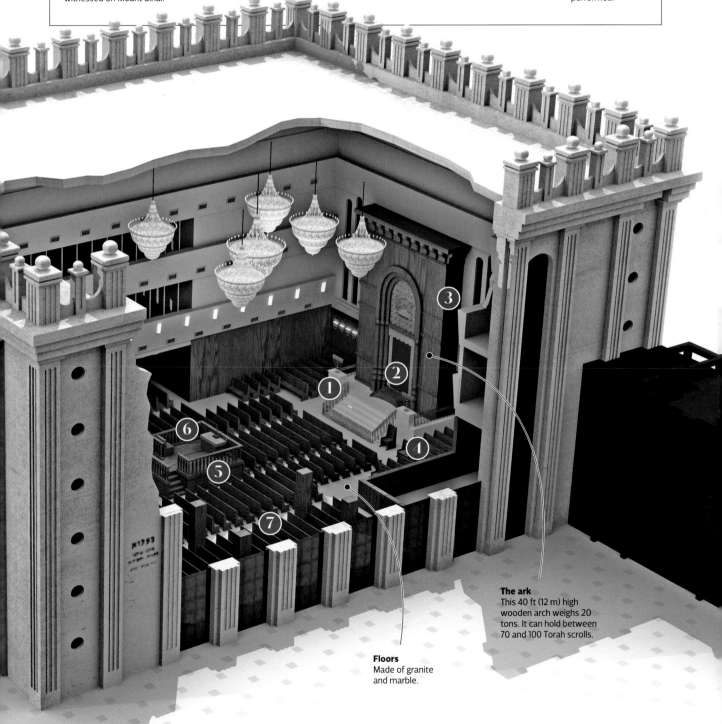

The ark
This 40 ft (12 m) high wooden arch weighs 20 tons. It can hold between 70 and 100 Torah scrolls.

Floors
Made of granite and marble.

Holy Sepulchre

The Holy Sepulchre is closely related to Catholic and Orthodox Christianity because according to the Gospels, this is the site of Christ's crucifixion, burial and resurrection. It is located in the Old City of Jerusalem between the East (Arab) and the West (Jewish). The Church of the Holy Sepulchre has become one of the most sacred sites in Christianity and has been the center of pilgrimages since the 4th century. Today it houses the Orthodox Patriarchy of Jerusalem.

Reconstruction

After Christianity was officially recognized in 326, the Emperor Constantine built the Church of the Holy Sepulchre on the spot of Christ's crucifixion and resurrection. The building was destroyed and rebuilt many times because of wars.

The main vault
Is in the Byzantine style, consisting of a half barrel vault built in slices and without centring. The first slice ensures the adhesion of the bricks to the head wall using mortar where the arch has previously been traced.

Monastic rooms
Were added to the Church away from the central worshipping areas. However, many of them have small private use chapels for the monks.

The monastery
Is an independent Coptic church since the Chalcedon Council in 451. It has preserved the Christian Orthodox belief and doctrine in the purest form.

THE SACRED SITES
To preserve the crucifixión and resurrection site, the Byzantines covered the Stone to set up the sanctuary keeping it elevated above Christ's tomb and the Golgotha hill where he was crucified.

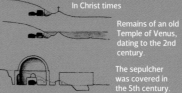

In Christ times

Remains of an old Temple of Venus, dating to the 2nd century.

The sepulcher was covered in the 5th century.

The lower walls
Are built of stone, but the upper walls are made of brick to reduce the weight. However, since brick is more fragile, small collapses can be common.

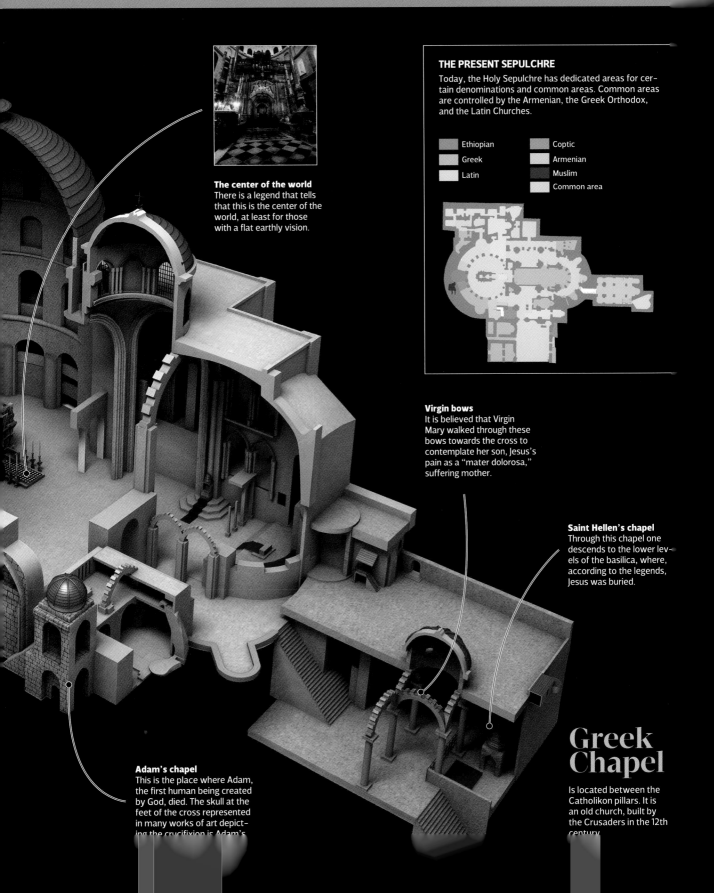

The center of the world
There is a legend that tells that this is the center of the world, at least for those with a flat earthly vision.

THE PRESENT SEPULCHRE
Today, the Holy Sepulchre has dedicated areas for certain denominations and common areas. Common areas are controlled by the Armenian, the Greek Orthodox, and the Latin Churches.

- Ethiopian
- Greek
- Latin
- Coptic
- Armenian
- Muslim
- Common area

Virgin bows
It is believed that Virgin Mary walked through these bows towards the cross to contemplate her son, Jesus's pain as a "mater dolorosa," suffering mother.

Saint Hellen's chapel
Through this chapel one descends to the lower levels of the basilica, where, according to the legends, Jesus was buried.

Greek Chapel
Is located between the Catholikon pillars. It is an old church, built by the Crusaders in the 12th century.

Adam's chapel
This is the place where Adam, the first human being created by God, died. The skull at the feet of the cross represented in many works of art depicting the crucifixion is Adam's

Sydney Opera House

This innovative building, one of the architectural icons of the 20th century, was designed by Danish Jørn Utzon. It was inaugurated in 1973, and designated a World Heritage Site in 2007. It houses the opera house, a concert hall, large and small theatres, an exhibition area and a library. The domes are aligned along a shared axis covering three buildings, and have been arranged on a platform.

Icon of Australia

The Sydney Opera House is the symbol of Australia and one of the most famous buildings in the world. Distantly, the building seems to float above the sea. ↘

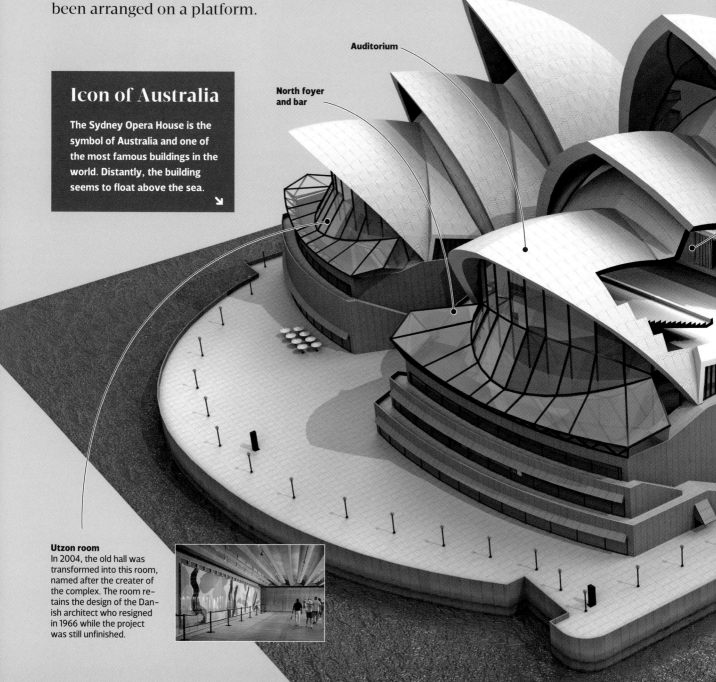

Auditorium

North foyer and bar

Utzon room
In 2004, the old hall was transformed into this room, named after the creator of the complex. The room retains the design of the Danish architect who resigned in 1966 while the project was still unfinished.

Concert hall
This is the largest of the five theatres of the Sydney Opera House. Symphonies, opera, ballet and theatre of great cultural significance are debuted here.

The vaults
The vaults are supported by a system of concrete ribs and covered in more than a million white and cream coloured tiles.

South foyer and bar

OLYMPIC OPERA HOUSE
During the 2000 Olympic Games held in Sydney, the triathlon swimming trials were held on the shores of the theatre, while the cycling and running events were held in the nearby Royal Botanic Gardens.

233

This was the number of ideas presented to build the Opera House. That of Danish architect Jørn Utzon was the eventual winner.

Restaurant
The restaurant of the renowned chef Guillaume Brahimi provides French haute cuisine and enviable views. The complex also has three other restaurants and six bars.

Stage

PLINTH
This solid base distributes the service spaces: dressing rooms, changing rooms, storerooms, offices and the library.

Wings

Stage

South foyer and bar

North foyer and bar

Auditorium

Dressing rooms

Storage area

Orchestra pit

Burj Al Arab

Built on an artificial island in the United Arab Emirates, the design is inspired by yachts, seafaring tradition and the old ships of Dubai. The structure of this marvel, one of the few 7-star hotels in the world, combines state-of-the-art technology, concrete, steel, glass and a gigantic glass fiber cloth. The 202 luxury suites of the Burj Al Arab look down over the Arabian Gulf from a height of 1,053 ft (321 m).

The giant sail

The hotel shape was inspired by a yacht and the structures that house the rooms are joined by a special fabric called Dyneon. They are made of fiberglass covered with Teflon to protect the structure from wind, the high temperaturas of the desert and dirt. It has a surface equivalent to 3.71 acres and is divided in twelve pieces. →

162,000 ft² (15,000 m²)

The dimensions of the building's gigantic glass fiber cloth, which is divided into twelve panels.

↑ LIGHTS
Strobe lights provide 150 colour changes as the night progresses. The lights are projected onto the giant sail which covers the building.

Cooling effect
Due to the area's high levels of sunlight and intense heat, the front section was covered in a special white double skin, which allows light to pass but reflects part of the sun's heat.

The suites
It has 202 luxury suites between 554 and 2,559 ft² (169 and 780 m²), equipped with advanced technology and distributed over two wings supported by a large steel structure.

↓ HELIPORT This is 695 ft (212 m) high. In February 2005, Roger Federer and André Agassi played a game of tennis as part of a publicity campaign.

The island
Supported by 250 columns which penetrate some 148 ft (45 m) below sea level, this triangular shaped island has sides measuring 492 ft (150 m).

Design
The island's edge has holes that break up the swell.

Main column
The mast, measuring 196 ft (60 m), is the top of a reinforced concrete spine which supports the steel structure.

Skeleton
The steel structure supports the rest of the building, built as two wings connected by a sail.

Heliport

RESTAURANTS

Of the nine restaurants, one—Al Muntaha—is located high up on a platform jutting out 89 ft (27 m) on each side. Another of the hotel's well known restaurants is Al Mahara, which is located below sea level and offers diners underwater views.

Under the sea. The tables of Al Mahara look onto a spectacular natural aquarium.

The atrium
At 597 ft (182 m) high, it is actually the tallest lobby in the world and is decorated with marble, gold leaf, velvet, dancing fountains, light shows and aquariums.

Access
The hotel is located 820 ft (250 m) off the coast and is connected by a bridge.

Beijing's National Stadium

For decades now, the enormous popularity and interest that major sporting events arouse, together with the significant commercial interests attached to them, has led to the construction of spectacular sporting venues. These multi-million dollar architectural jewels are equipped with the latest technology at all levels. A clear example is the Beijing National Stadium, the main venue for the 2008 Olympic Games. Every four years the Olympic host nation makes every effort to build the best stadium possible.

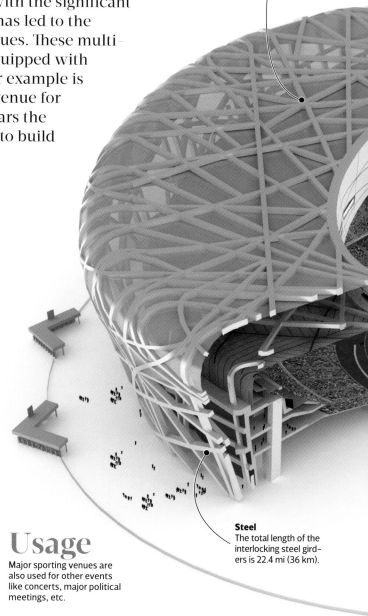

Construction
About 17,000 people worked on the stadium.

Steel
The total length of the interlocking steel girders is 22.4 mi (36 km).

OTHER STADIUMS

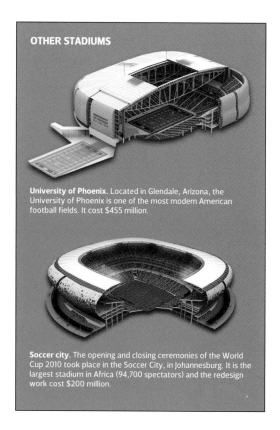

University of Phoenix. Located in Glendale, Arizona, the University of Phoenix is one of the most modern American football fields. It cost $455 million.

Soccer city. The opening and closing ceremonies of the World Cup 2010 took place in the Soccer City, in Johannesburg. It is the largest stadium in Africa (94,700 spectators) and the redesign work cost $200 million.

Usage

Major sporting venues are also used for other events like concerts, major political meetings, etc.

BIRD'S NEST

It is 1082 ft (330 m) long, 971 ft (296 m) wide and 226 ft (69 m) high. It has a capacity for 80,000 spectators.

Membrane
The roof is covered with a transparent double membrane.

Olympic Stadium

Known as "The Bird's Nest" for its appearance, it was designed by the Swiss firm Herzog and de Meuron. It cost $420 million and was built especially for the Beijing Games. It took four and a half years to complete it.

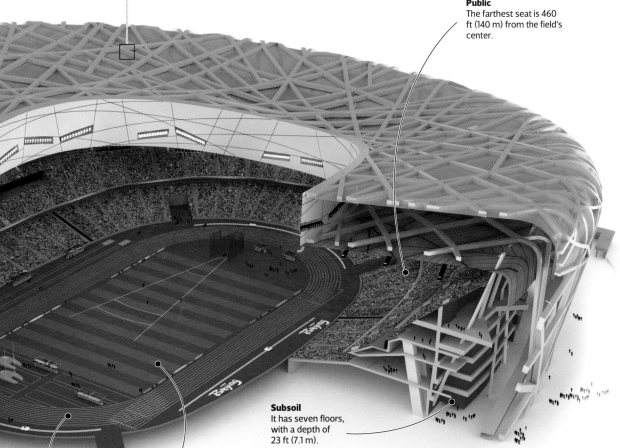

Public
The farthest seat is 460 ft (140 m) from the field's center.

Subsoil
It has seven floors, with a depth of 23 ft (7.1 m).

Playground
The playing área for sports is 25,590 ft² (7,800 m²).

Athletics track
It has nine running tracks with 1312 ft (400 m) each in length.

FLOOR AREA: 160.3 MI² (258 KM²)

ENERGY SAVING

The ETFE (high strengh plastic) panels allow the passage of light and save energy.

Daylight 100%

18.4%

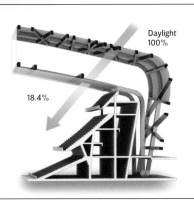

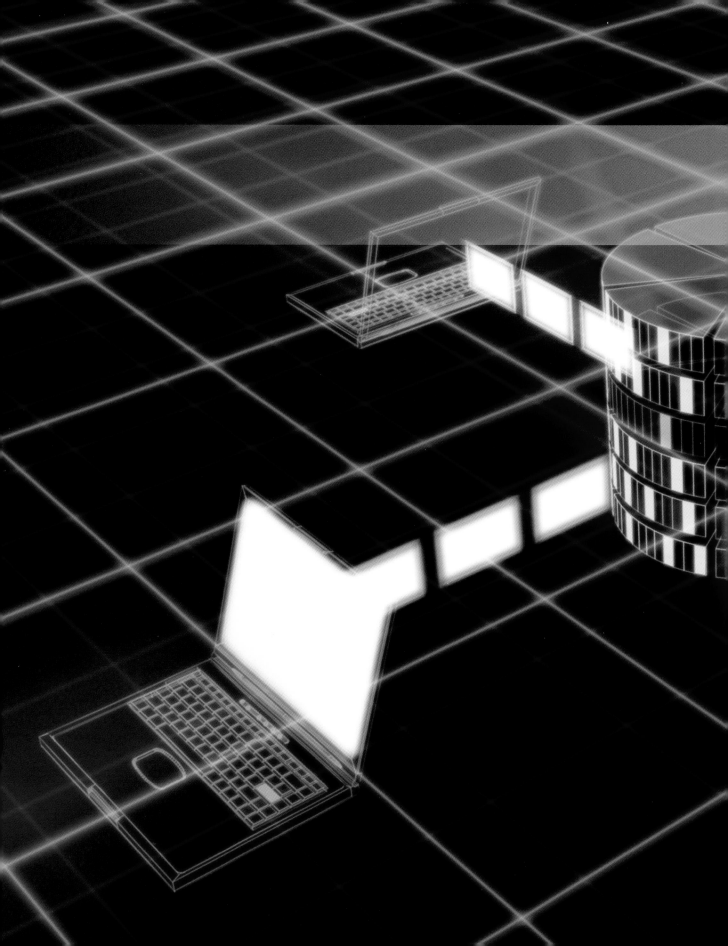

COMMUNICATION

The electrical telegraph

The telegraph changed not only communications technology, but the way of life for societies around the world. Today, the telegraph is a museum piece, but when it was invented in the early nineteenth century it caused a technological revolution. For the first time, messages could be sent and received in real time and across long distances. Telegraphs had low operating costs, but high construction costs, at least at first. When it was invented, messages could only be sent between two points physically connected by a wire. Over time, the chord was cut and telegraphs could be sent without being connected.

Battery
It stores the energy to produce electricity when the pulser is driven.

Points and rays

Although numerous telegraphs designs existed, the basic concept always stayed intact. A telegraph sends electrical signals to the other side of the line when the operator drives the pulser and closes the electrical circuit. An electromagnet then drives a striker pin according to the signals, marking a paper tape in the receiving station. ↘

Pulser
It is driven by the operator. When pushed down, an electrical current is completed in the cable. A brief pulse is a dot; a long pulse is a dash.

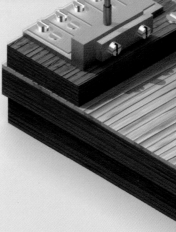

↑ THE TELETYPE The teletype was an advanced form of telegraphy, common in the 20th century. Teletype machines did not have to be connected by wires and could receive several messages simultaneously. Some models even had a monitor!

1874

Thomas Alva Edison patents the telegraph duplex. This allowed users to transmit more than one message simultaneously.

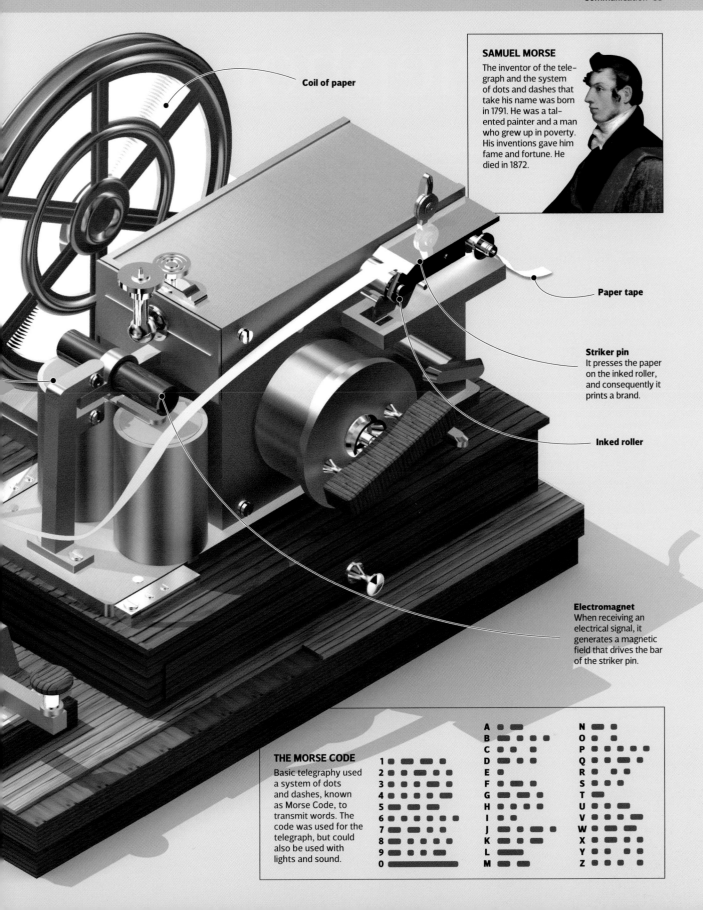

Coil of paper

SAMUEL MORSE
The inventor of the tele-graph and the system of dots and dashes that take his name was born in 1791. He was a tal-ented painter and a man who grew up in poverty. His inventions gave him fame and fortune. He died in 1872.

Paper tape

Striker pin
It presses the paper on the inked roller, and consequently it prints a brand.

Inked roller

Electromagnet
When receiving an electrical signal, it generates a magnetic field that drives the bar of the striker pin.

THE MORSE CODE
Basic telegraphy used a system of dots and dashes, known as Morse Code, to transmit words. The code was used for the telegraph, but could also be used with lights and sound.

The telephone

Without phones, the world as we know it today would simply not exist. The phone is a ubiquitous element in all both homes and offices. Since the advent of the mobile phone, it has also been in people's pockets. The phone developed in the 1870s and completely changed the way people communicated. The phone narrowed distances, allowing for direct communication across vast geographic areas.

Electronic circuits
They fit and amplify the signals to facilitate hearing.

A simple concept

The sound of the speaker's voice becomes electricity, travels through the line, and returns to sound in the ear of the receiver. Technological advances since the telephones' development have improved it. But the basic concept stays intact.

→

Switch
When picking up the telephone, it activates the contact with the local telephone exchange to connect a call.

Earpiece
It vibrates in agreement with the electrical current to recreate the sound that arrives from the other side of the line.

DIFFERENT WAYS OF COMMUNICATING

Telephone calls follow different paths, based on the telephone used and the distance covered.

The fixed telephone
These telephones are connected through mesh wiring, such as copper, or more recently, optical fiber. Sometimes, the communication includes the satellite.

The mobile telephone
It's connected with ground antennas of limited range. These mobile phone towers work together, passing calls to different towers if the caller leaves range.

The satelite telephone
It receives the communication directly from a satellite, which is why it allows calls from anywhere on the planet. That is also why it is so expensive.

↓ **EVOLUTION** Since its original conception in the nineteenth century, the phone has been experiencing substantial improvements to become the powerful communication device we know today.

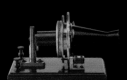

1876
Alexander Graham Bell patents the telephone. It allows for communication across short distances.

1880s
The first hand cranked wall telephones.

1905
Almon Strowger invented the first dial telephone. The operators were no longer needed.

1937
The first telephones with a ringer inside the box of the device. The previous ones had it outside.

1963
Push-bottons telephones replace the rotary dial models.

1963

Touch-Tone dialing is released to the general public, simplifying telephone technology for the average user.

Cable
It sends and receives the sounds, turned into electrical waves.

Alphanumeric keyboard
Each key, when pressed, sends a combination of two tones at different frequencies. These two tones represent a particular number.

Microphone
It vibrates in agreement with the voice and it transforms these vibrations into electrical waves, that are soon amplified on the telephone.

TONES AND PULSES

When the telephone was first invented, dialing was done by rotary dialing. By the 1980s, touch-tone dialing was used nearly everywhere.

Rotary dialing
It depends on a mechanical device that indicates to the telephone exchange the amount of rotation for each number. Rotary dialing was used from the time of the phone's invention.

Touch-tone dialing
Instead of electrical pulses, each number emits two sounds in different frequencies that the power station interprets. Compared to rotary dialing, it has less errors and greater speed.

1970s
Cordless phones reach the general public

1983
The first mobile phone is commercialized: Motorola DynaTAC 8000x

1990
The mobiles reduce their size to that of a candy bar and increase the capabilities. First text message (1992)

2003
BlackBerry with phone, texting, and email arrives on the market

2007
iPhone 7 inaugurates the era of smartphones: devices with multiple applications and Internet access

2018
Smartphones have great speed of connection to networks (5G)

Television set

Only thirty years after the widespread adoption of radio technology, television again revolutionized society. While the technology existed by the end of the nineteenth century, television only gained popularity among the general public after the Second World War. Today, television remains one of the most influential mass media. It is both loved and hated, praised and criticized, and is considered both a an invaluable tool for advertising and for informing. Recent advances in technology have allowed for televisions to become both larger and thinner.

Behind the screen

A TV receiver is a device able to turn radio-electric signals into sound and images. In order to display the image in colors, the traditional method used in televisions was to have phosphorus compounds that ignite when receiving an electrical current.

→

Julius Plücker

Mathematician and German physicist, Plücker was a pioneer in the investigation of cathode rays, which, years later, made possible the development of the television.

THE SCREEN

Is constructed of phosphorus. This element ignites when it receives an electron stream. Each beam can ignite a point of a certain color.

This image shows a part of a screen, magnified many times. Human eyes are incapable of differentiating the points and it is possible to create any color by varying the intensity of each point. The intensity of each point is determined by the intensity of the electrical signal that arrives at the television via the antenna.

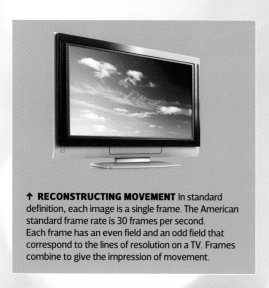

↑ **RECONSTRUCTING MOVEMENT** In standard definition, each image is a single frame. The American standard frame rate is 30 frames per second. Each frame has an even field and an odd field that correspond to the lines of resolution on a TV. Frames combine to give the impression of movement.

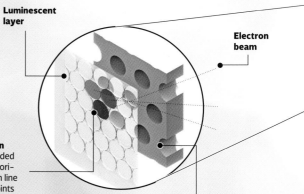

Luminescent layer

Electron beam

Red, blue, green
The image is divided into a series of horizontal lines. Each line is made up of points of different brightness. By convention, they are split into the three primary colors: red, green, and blue.

Shadow mask
This is a metallic layer, which filters electron beams, "separating" some from the others.

THE SIGNAL'S PATH

There are two basic forms of transmission for traditional TV: by antenna or by cable.

By antenna
An electromagnetic signal composed of high frequency waves is emitted. As the reach is small, repeaters or liaison stations are used.

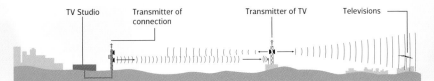

TV Studio

Transmitter of connection

Transmitter of TV

Televisions

By cable
The signal arrives by a cable. The infrastructure for this system is more expensive, but the image quality is better.

Electron gun
It generates three types of electron beams that glow in the three primary colors: blue, red and green. The intensity of each depends on the signal that comes from the antenna.

Electron beam

Sound

Is encoded and broadcast using the same method as the one employed in FM radio.

Deflection coils
They generate a magnetic field that organizes the electrons before they reach the screen.

The antenna
A classic TV antenna has different size dipoles to facilitate the reception from signals of different frequencies. TVs can also receive signals via cable.

The fax

Fax machines appeared in the 1980s, taking over corporate society. In essence, the fax could be considered the direct predecessor of e‑mail, since, for the first time, text could be sent immediately from place to place. With some technical difficulties, images could also be transmitted. Fax machines work by dividing a sheet into a grid and transmitting line‑by‑line, from top‑to bottom. Dark and clear spaces are received by the intended unit and transcribed. Despite other technological advances, the fax machine continues to have a spot in offices across the globe.

Three in one

The devices for sending and receiving faxes consists of three integrated parts: a phone, a scanner and a printer.

→

1985

The GammaFax, the first personal computer fax board, is released.

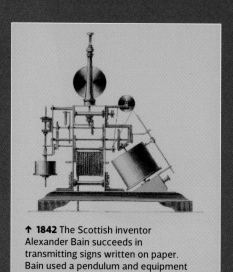

↑ 1842 The Scottish inventor Alexander Bain succeeds in transmitting signs written on paper. Bain used a pendulum and equipment of his own invention.

The scanner divides the page into lines, and for each line detects the dark points and the clear points.

Here you can select se
according to the type

Each part can have 0.008 in (0.2 mm) of height.

W B W W W B W B

The result is a set of white and black points. This process is repeated for every line until the end of the document.

The dark points and the clear points become electrical impulses, that are interpreted by re‑ceiving fax machine as it prints the message.

Low tones (white)   High tones (black)

The printer interprets the electrical pulses and places points or blanks in the corresponding places.

Button keyboard
Here the user can select several op‑tions, according to the type of device.

Earpiece
It allows the fax machine to be used as a common telephone.

Print tray for sent faxes

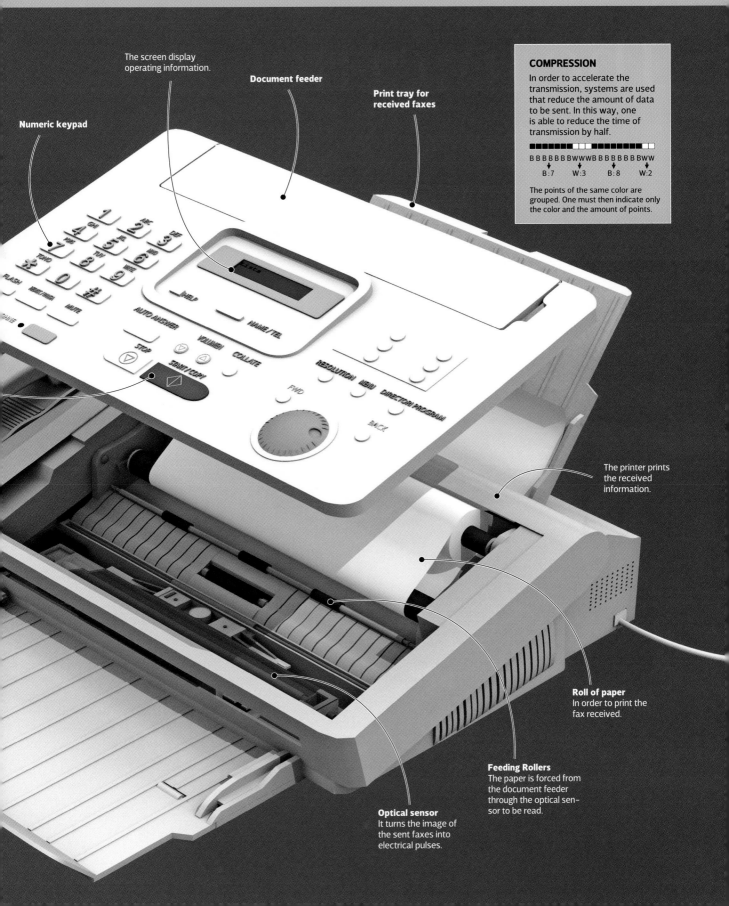

The screen display operating information.

Document feeder

Print tray for received faxes

Numeric keypad

COMPRESSION
In order to accelerate the transmission, systems are used that reduce the amount of data to be sent. In this way, one is able to reduce the time of transmission by half.

B B B B B B B W W W B B B B B B B W W
B:7   W:3   B:8   W:2

The points of the same color are grouped. One must then indicate only the color and the amount of points.

The printer prints the received information.

Roll of paper
In order to print the fax received.

Feeding Rollers
The paper is forced from the document feeder through the optical sensor to be read.

Optical sensor
It turns the image of the sent faxes into electrical pulses.

The Internet

The Internet is a network that connects all of the computers in the planet. The Internet allows users to communicate with the world in different ways — to see and talk directly with people who are far away, to shop, without leaving the house, to send and receive mail, read the newspapers, make financial transactions, and more. The Internet is accessible to all computers that are connected to the Web.

Around the world

Information on the Internet travels fragmented into small packets of data and using different routes. Packages (a Web page, an image, etc.) are reassembled when they arrive at the destination.

① COMPUTER
The computer, or terminal, is assigned an IP number to give it a specific identity. With that number, the computer makes a request to the Internet Service Provider's (ISP) server. The request contains another IP number that represents the identification of the required information.

② ISP SERVER
ISP servers provide computers with networks of internet connection. They also host websites. When they receive a request from a computer, they send the order over networks around the world.

The packages
They are fragments of binary information (a language of ones and zeros) that are part of a whole. Each packet is labeled with numbers that represent the source and destination address.

Computer

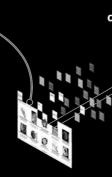

ISP server

↑ **A WEBSITE OR WEB PAGE** Contains a series of documents written in hypertext markup language (HTML) combined with other, more sophisticated languages, such as Java and Flash animation.

Protocol
A protocol is a set of rules and conventions used to connect and exchange information between computers and servers around the world.

Global Connections

It is estimated that there are about 3 billion Internet users, almost half of the world's population.

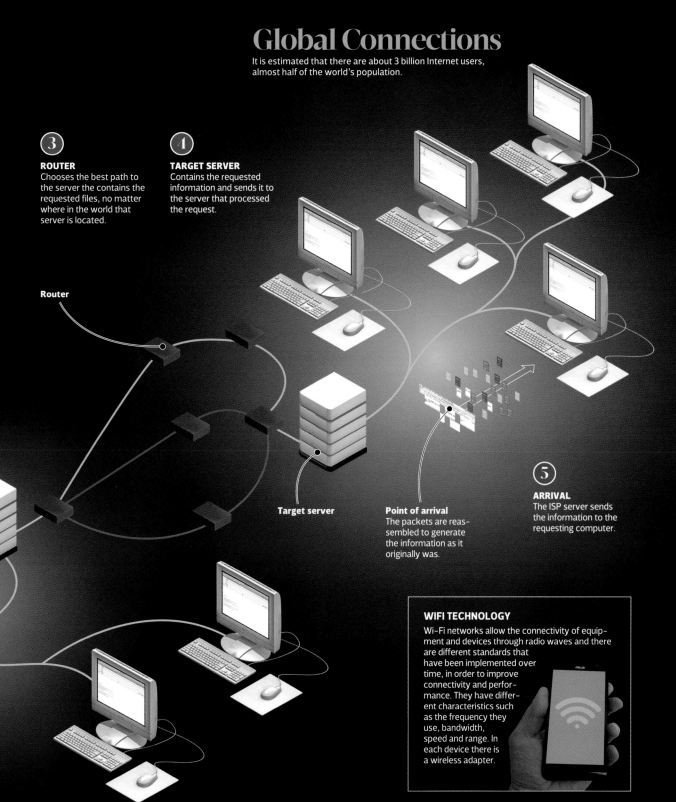

③ **ROUTER**
Chooses the best path to the server the contains the requested files, no matter where in the world that server is located.

④ **TARGET SERVER**
Contains the requested information and sends it to the server that processed the request.

Router

Target server

Point of arrival
The packets are reassembled to generate the information as it originally was.

⑤ **ARRIVAL**
The ISP server sends the information to the requesting computer.

WIFI TECHNOLOGY

Wi-Fi networks allow the connectivity of equipment and devices through radio waves and there are different standards that have been implemented over time, in order to improve connectivity and performance. They have different characteristics such as the frequency they use, bandwidth, speed and range. In each device there is a wireless adapter.

E-mail

Just two decades ago, sending a letter meant waiting days, weeks, or months for it to be delivered. The telegram provided an alternative for short messages, and the phone allowed direct communication but lacked a paper trail. Fax machines provided another alternative, but at high cost. The development of e mail changed everything. In addition to text, emails can contain music, videos, images, spreadsheets, and more, and are very efficient, fast, and cheap. For these reasons, e mail has displaced ordinary mail for some uses.

A complex process

The fraction of a second that it takes for an e-mail message to travel from the sender to the recipient hides a complex mechanism with twists and turns. ↘

1982

The protocol SMTP begins to be used and e-mail becomes known worldwide as such.

③ **DNS SERVER**
The "expert" address book, the DNS server translates the domain to which the e-mail is directed into an IP address and determines the appropriate mailbox (MX server). It responds to the SMPT server with that data.

② **SMPT SERVER**
Like a post office, the SMPT server contacted by the client receives the e-mail and sends a query to the DNS server.

① **SENDER**
A user sends an e-mail from a computer. The message travels to the mail server, known as an SMTP server.

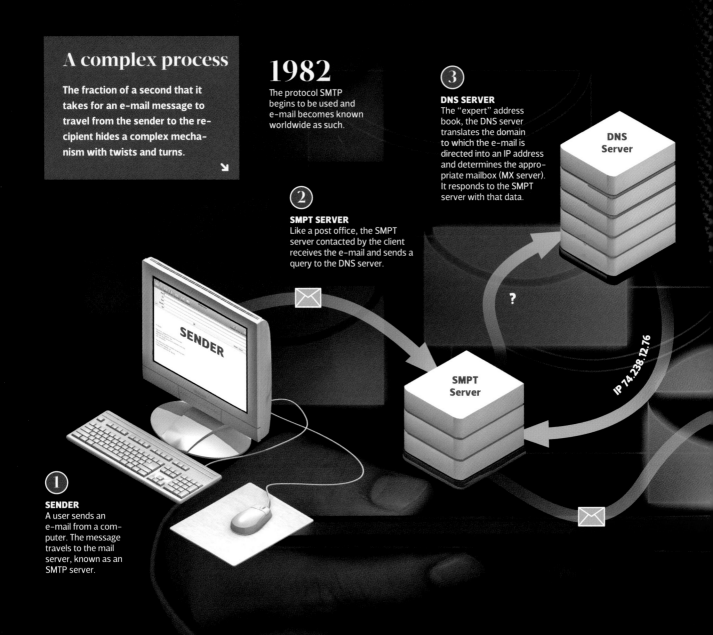

DNS Server

SMPT Server

IP 74.238.12.76

SENDER

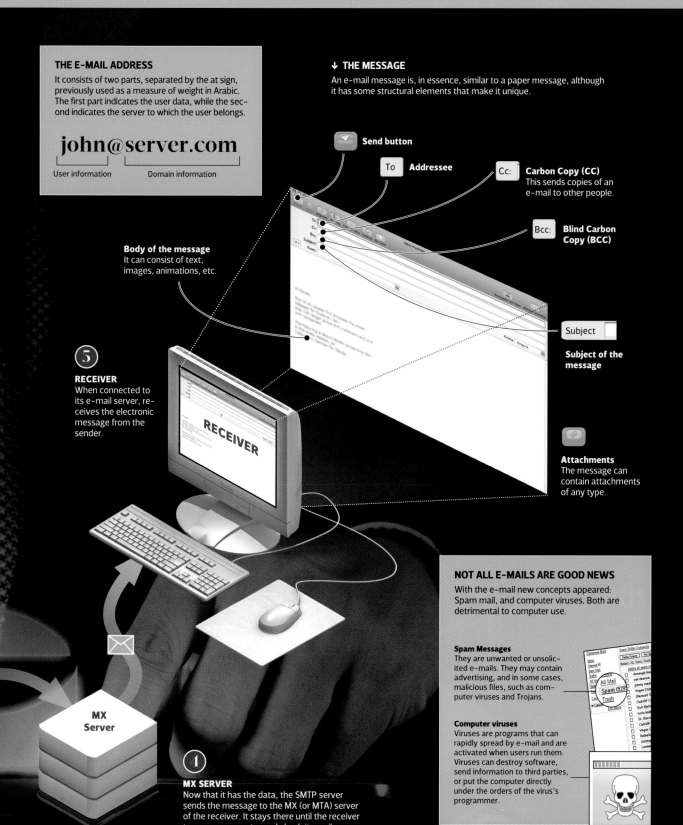

THE E-MAIL ADDRESS

It consists of two parts, separated by the at sign, previously used as a measure of weight in Arabic. The first part indicates the user data, while the second indicates the server to which the user belongs.

john@server.com

User information | Domain information

↓ THE MESSAGE

An e-mail message is, in essence, similar to a paper message, although it has some structural elements that make it unique.

Send button

To **Addressee**

Cc: Carbon Copy (CC)
This sends copies of an e-mail to other people.

Bcc: Blind Carbon Copy (BCC)

Body of the message
It can consist of text, images, animations, etc.

Subject

Subject of the message

⑤ **RECEIVER**
When connected to its e-mail server, receives the electronic message from the sender.

Attachments
The message can contain attachments of any type.

NOT ALL E-MAILS ARE GOOD NEWS

With the e-mail new concepts appeared: Spam mail, and computer viruses. Both are detrimental to computer use.

Spam Messages
They are unwanted or unsolicited e-mails. They may contain advertising, and in some cases, malicious files, such as computer viruses and Trojans.

Computer viruses
Viruses are programs that can rapidly spread by e-mail and are activated when users run them. Viruses can destroy software, send information to third parties, or put the computer directly under the orders of the virus's programmer.

MX Server

④ **MX SERVER**
Now that it has the data, the SMTP server sends the message to the MX (or MTA) server of the receiver. It stays there until the receiver

Digital TV

The invention of liquid crystal display (LCD) televisions, especially high definition televisions, caused picture quality to rise to new heights. However, a concurrent development made this revolution possible – digital television. Digital TV has replaced the traditional analog television system. The concept involves the digitization of the entire television process, from capturing the images in a complex series of zeroes and ones, to the conversion from the ones and zeroes to colored pixels.

A digital world

Digital TV, like analog, begins in a studio. It is generally transmitted by cable or satellite to every TV capable of receiving its signal.

① IMAGE
High resolution digital cameras take video and encode the information in binary, a language of ones and zeroes.

② STORAGE
The information is transmitted to computers working at high-speeds to transfer it to online servers.

③ TRANSMISSION
Performed in compressed digital formats allowing one to send huge flows of information at very high speeds, whether by cable or by satellite.

LCDs

Liquid crystal display technology has been applied to television sets, causing a revolution in size and image quality. LCD televisions are flat-screen and lighter than conventional sets and need less power to operate.

2009

The year in which all TV stations in the United States were required to broadcast digitally.

→ MULTIPLE FORMATS, MULTIPLE CHANNELS
One of the most important characteristics of digital TV is the ability of TV stations to subdivide their signal into channels in lower resolution with different programming, or to broadcast a single channel in the best resolution, due to the high speed of data transmission.

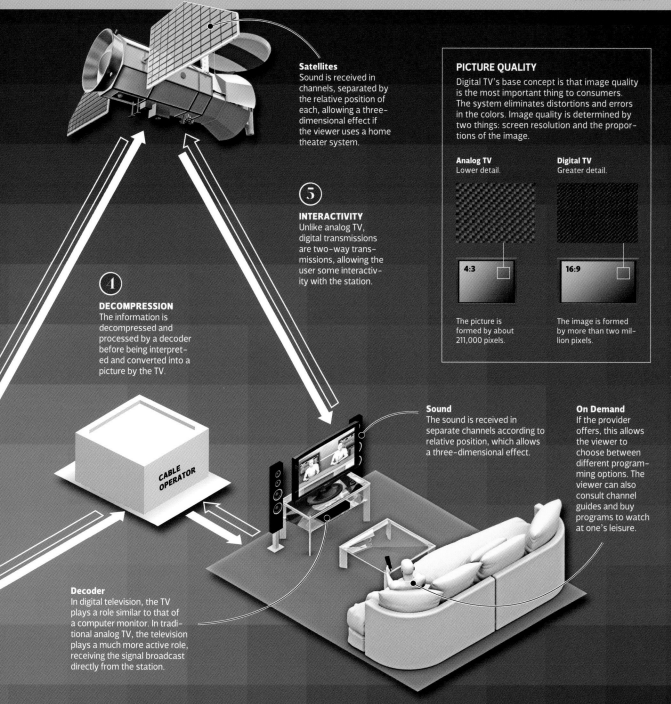

Satellites
Sound is received in channels, separated by the relative position of each, allowing a three-dimensional effect if the viewer uses a home theater system.

⑤

INTERACTIVITY
Unlike analog TV, digital transmissions are two-way transmissions, allowing the user some interactivity with the station.

PICTURE QUALITY

Digital TV's base concept is that image quality is the most important thing to consumers. The system eliminates distortions and errors in the colors. Image quality is determined by two things: screen resolution and the proportions of the image.

Analog TV
Lower detail.

Digital TV
Greater detail.

4:3

16:9

The picture is formed by about 211,000 pixels.

The image is formed by more than two million pixels.

④

DECOMPRESSION
The information is decompressed and processed by a decoder before being interpreted and converted into a picture by the TV.

Sound
The sound is received in separate channels according to relative position, which allows a three-dimensional effect.

On Demand
If the provider offers, this allows the viewer to choose between different programming options. The viewer can also consult channel guides and buy programs to watch at one's leisure.

CABLE OPERATOR

Decoder
In digital television, the TV plays a role similar to that of a computer monitor. In traditional analog TV, the television plays a much more active role, receiving the signal broadcast directly from the station.

DIFFERENT FORMATS

	Analog	SDTV	EDTV	HDTV	HDTV
Pixels	211,000	307,200	337,920	921,600	2,073,600
Resolution	640x480	640x480	704x480	1280x720	1290x1080
Format scanning	480 lines	480 lines	480 pixels	720 pixels	1080 lines
Screen	4:3	4:3	4:3 o 16:9	16:9	16:9
Quality	Common	Good	Very good	Excellent	Excellent

Satellites

Much of the communication technology that society relies on – the telephone, the radio, and the television – existed by the 1950s. Underwater phone cables, though limited in capacity, connected the continents. Some radio signals were also strong enough to cross the oceans, without the need for cables. For television, however, it was impossible for broadcasts to cross the ocean and reach global audiences. In 1962, the launch of the first telecommunications satellite changed all of that. For the first time, the planet was truly connected.

True revolution

A satellite can receive a signal emitted from the Earth (a telephone conversation, for example) and then relay it to another point on the planet. Communications satellites are placed in geosynchronous orbit, that is to say, that turn at the same speed as the Earth, and therefore, remain over the same point of the planet. Satellites are typically 22,369 mi (36,000 km) from the surface.

TRANSMISSION VIA SATELLITE

The principle of transmitting via satellite is very simple since the radio waves travel at the speed of light. There-fore, the connection between two far-removed points of the planet is almost immediate.

①
A station on Earth transmits the signal toward the satellite.

②
The satellite receives and amplifies the signal. It then relays it.

③
The receiving station on Earth re-ceives the satellite's signal and relays it at local level.

MODERN SATELLITES

Fixed sending and receiving antenna. Can target specific places on Earth.

Transponder. This is the heart of the satellite. It corrects for atmosphere-produced distortions of radio signals.

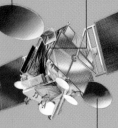

Reflector. Captures signals and retrans-mits them directly

Solar panels. Take the light from the Sun and transform it into electrical energy.

The antennas
In order to send and receive signals from the satellite antennas weighing 380 tons were built in buildings 15 floors tall.

Andover, Maine, USA ❶

→ THE FIRST TRANSMISSION
July 11, 1962 saw the first satellite transmission in history. An image of an American flag flying in Andover, Maine, USA, was broadcast to Pleumeur–Bodou, France. The first public transmission was twelve days later.

22,400 mi
(36,000 km)

This is the required distance of a satellite's orbit so that it remains fixed with respect to Earth.

Antenna
It is used to issue orders to the satellite from Earth. Simultaneously, the satellite issues data on its position and its operating capacity.

TELSTAR I

Telstar I capacity
600 telephone channels or one television channel.

Equatorial antennas
They receive microwave signals from Earth, that are amplified.

Amplifiers TWT
They amplify the signal from Earth to be relayed.

Solar panels
They turn the Sun's light into electrical energy that feeds the satellite.

③ Pleumeur-Bo-dou, France

THE ORBIT
Satellites use different orbits, depending on their function.

Geosynchronous orbit
Present communica-tions satellites are in orbit 22,400 mi (36,000 km) from Earth. They orbit at exactly the same speed, so they are always on a fixed terrestrial point.

Elliptical orbit
Telstar 1 orbited the planet elliptically, making a complete revolution every 2.5 hours. As a result, satellite was only able to transmit live for approximately 20 minutes.

GPS

The Global Positioning System (GPS) was developed by the US Department of Defense. It makes it possible to establish the position of a person, vehicle, or aircraft anywhere in the world. To do this, it employs a system of two dozen NAVSTAR satellites. GPS was fully deployed in 1995 and, although it was designed for military use, is now a necessary part of daily life for many people. The European Union is developing a similar system, called Galileo, that will consist of 30 satellites.

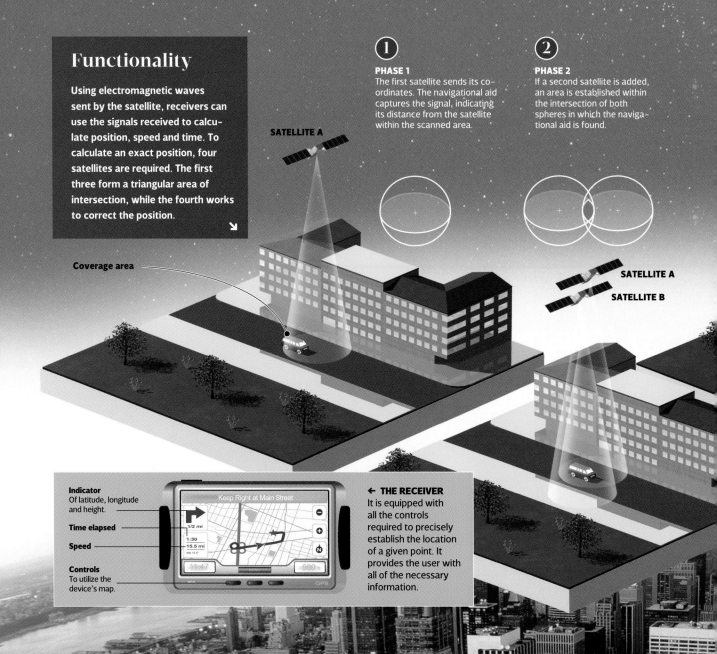

Functionality

Using electromagnetic waves sent by the satellite, receivers can use the signals received to calculate position, speed and time. To calculate an exact position, four satellites are required. The first three form a triangular area of intersection, while the fourth works to correct the position.

↘

Coverage area

① PHASE 1
The first satellite sends its co-ordinates. The navigational aid captures the signal, indicating its distance from the satellite within the scanned area.

② PHASE 2
If a second satellite is added, an area is established within the intersection of both spheres in which the navigational aid is found.

SATELLITE A

SATELLITE A
SATELLITE B

Indicator
Of latitude, longitude and height.

Time elapsed

Speed

Controls
To utilize the device's map.

Keep Right at Main Street
1/2 mi
1:30
15.5 mi

← THE RECEIVER
It is equipped with all the controls required to precisely establish the location of a given point. It provides the user with all of the necessary information.

APPLICATIONS

Although it was originally developed as a navigational system, GPS is used in a variety of fields. The free use of this tool for work, business, recreation, and sports activities is changing the way we move and act.

Sports
GPS devices keep the athlete informed of time, speed, and distance.

Military
Used in remote-controlled and navigational systems.

Scientific
Used in paleontology, archaeology, and animal tracking.

Exploration
Provides orientation and marks reference points.

Transportation
Air and maritime navigation. Its use is growing in automobiles.

Agriculture
Maps areas of greater or lesser fertility within different plots of land.

$750

The annual cost, in millions, maintain the entire Global Positioning System.

③

PHASE 3
Combining the three satellites, a common point can be established that indicates the exact position of the navigational aid.

Waves

Using the electromagnetic waves sent by satellites, the receiver calculates the distance and position of the point sought. The waves travel at 186,400 mi/sec (300,000 km/sec).

④

PHASE 4
A fourth satellite is required to correct any possible positional error.

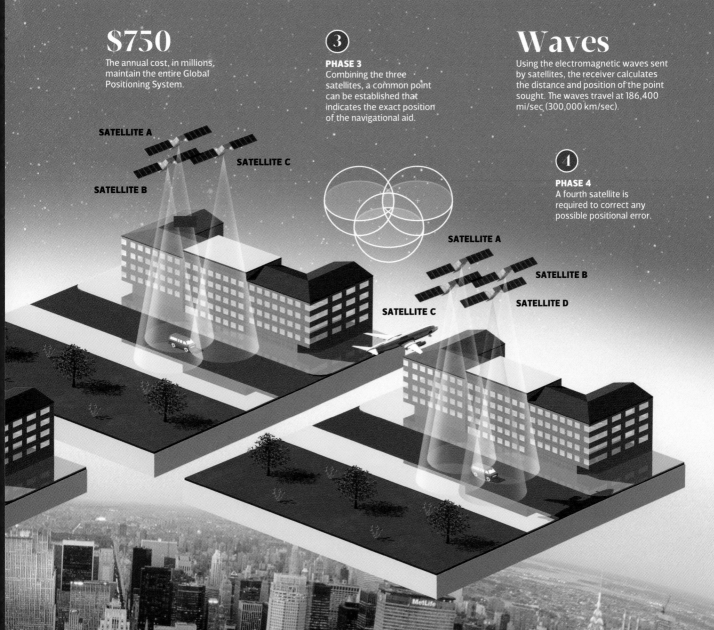

SATELLITE A

SATELLITE C

SATELLITE B

SATELLITE A

SATELLITE B

SATELLITE C

SATELLITE D

ENERGY

The steam engine

The external combustion engine, which transforms the energy of water vapor into mechanical energy, was essential to the Industrial Revolution that took place in England in the 17th and 18th centuries. Its invention is a long story that begins with rudimentary and impractical devices and continues up to the invention of the steam engine by James Watt. The steam engine was of fundamental importance for advancing industry and transportation, because it replaced beasts of burden, the mill, and even human laborers.

Transport
Watt's steam engine laid the groundwork for the use of high-pressure steam, a technique which was perfected during the 19th century and ushered in the application of the steam engine to the transport sector.

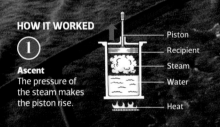

↑ **WATT'S GREAT CONTRIBUTION** James Watts was born in Scotland in 1736 and studied instrument making at the University of Glasgow. There, in 1763, he began to perfect the engine created by Thomas Newcomen in 1712, incorporating a separate chamber to condense steam. In 1769 he patented this innovation and laid the groundwork for its industrial application.

HOW IT WORKED

① **Ascent**
The pressure of the steam makes the piston rise.

Piston
Recipient
Steam
Water
Heat

Steam condenses

Water returns to its initial level

② **Descent**
Gravity makes the piston fall.

WATT'S INNOVATION

He placed a separate recipient where the steam was being condensed.

The valves let the steam pass from above or from below.

The piston rises or falls depending on the influx of steam.

Entrance

Exhaust

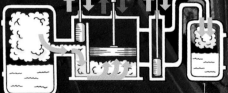

The steam ejected by the movement of the piston turns to liquid in the condenser.

Boiler

A revolutionary invention

By definition, the steam engine functions through the elastic force of boiling water. The pressure is used as a motivating force with the aid of mechanical devices, like the piston located in cylinders of prototypes produced by Newcomen and Watt. ↓

APPLICATIONS OF THE AGE

Mainly used in industry, mining and transportation.

Water extraction
Beginning with the previous model, in 1698, Thomas Savery patented a steam engine which was used to extract water from mines. Later, in 1712, Newcomen perfected it.

Spinning mill and weaving
Used primarily to operate machines for spinning and weaving and later, in the making of presses.

Sterilization
Around 1900 this model was constructed. It served, among other things, to sterilize water for nursing and to develop medicine.

Transportation
In boats, automobiles and locomotives.

Generation of electricity
Currently, this is one its most important uses. Steam passes throught a paddle wheel and its mechanical energy is converted to electrical energy.

289

Steam engines that were sold by the company Boulton & Watt in 1800, after 15 years of activity. The majority of machines were used in the textile industry and to drive water pumps and hammers in coal mines.

COMPARING SOURCES OF ENERGY
Around 1800

Watt's steam engine = **14 to 40 horses**
11 to 30 kilowatts

Horse moving a mill = **36.6 men** (minimum)
300 to 450 watts

Natural gas extraction

After petroleum, natural gas is the second most important energy resource on Earth because of its availability and efficiency. Natural gas has the reputation of being the cleanest fossil fuel. Technological advances, especially regarding the discovery of deposits, have produced an explosion in the reserve statistics over the last 15 years. These developments have led to natural gas being increasingly relied on to provide energy over large parts of the planet.

Phantom Energy

Natural gas is a colorless, odorless fluid that contains between 70 and 90 percent methane, the component that makes it useful as a source of energy.

LPG

Liquefied petroleum gas (LPG) is a byproduct of natural gas. It is bottled in cylinders and used by people who live in remote areas to operate, for instance, boilers and motors.

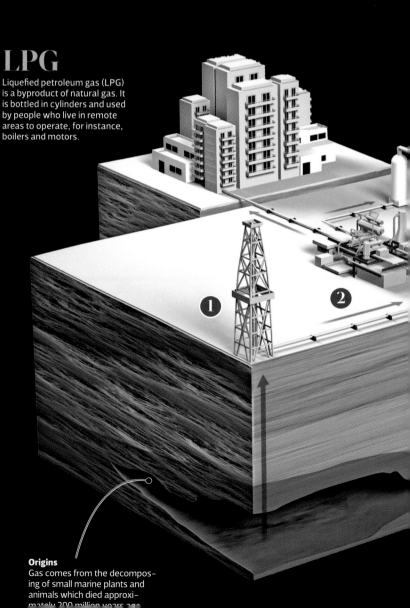

↑ **LOSSLESS TRIP** Among the many virtues of natural gas is the efficiency with which it can be transported. From gas deposits, it can be sent thousands of miles by ship or through gas pipelines with minimal losses.

Origins
Gas comes from the decomposing of small marine plants and animals which died approximately 200 million years ago.

EXTRACTION
The gas is extracted from the deposit through a hole. When the gas is under pressure, it rises to the surface on its own. When it is not under pressure, it must be pumped.

REFINEMENT
The solid and wet components are separated. Then the byproducts, like propane and ethylene, are separated.

DISTRIBUTION
After being distilled and converted, essentially into methane, natural gas is distributed for use through gas pipelines.

LIQUEFACTION
When it must be transported by sea or stored, the gas is compressed and cooled to −258° F (−161° C) to liquefy it.

TRANSPORTATION
Large, double-hulled, pressurized ships transport the gas in a liquid state.

GASIFICATION
After transport, the liquefied gas is returned to a gaseous state to be distributed through a network of gas mains.

DISTRIBUTION
The gas reaches residential and commercial consumers.

Transport without loss
Natural gas can be transported from its source without the risk of loss. It is transported by boat and gas ducts.

600

Is the number of times that the volume of gas decreases when it is liquefied for storage or transport.

DEPOSIT
Gas tends to be located inside porous rocks capped by impermeable rocks that are not necessarily associated with petroleum.

Dry gas deposits

Impermeable rock

Gas chamber

Impermeable rock

Petroleum deposits

Impermeable rock

Gas chamber

Petroleum

Petroleum extraction

Petroleum is the main energy source in the developed world. It comes from ancient organic deposits that have been buried in the depths of the Earth for hundreds of millions of years. Its pure state, called crude oil, is a mixture of different hydrocarbons of little use, and hence the oil must first be distilled to separate its components. This valuable resource, which pollutes the atmosphere when burned, is nonrenewable and available only in limited reserves; these characteristics have driven scientists to look for alternative energy sources.

Gas flare stack

② CRUDE OIL STORAGE
The crude oil is stored and then transported to refin-eries through pipelines or by large tanker ships.

2070
The year the world's oil reserves could run out if the current rate of consumption is maintained and no new discoveries are made.

③ VAPORIZATION
The crude oil is heated in a boiler up to 752° F (400° C) or more. Once vaporized, it is sent through the distilling tower.

① EXTRACTION
The oil is pumped from the deposit up to the storage tanks.

The process

After its extraction, crude oil is distilled and fractioned into several products, among them gasoline. ↑

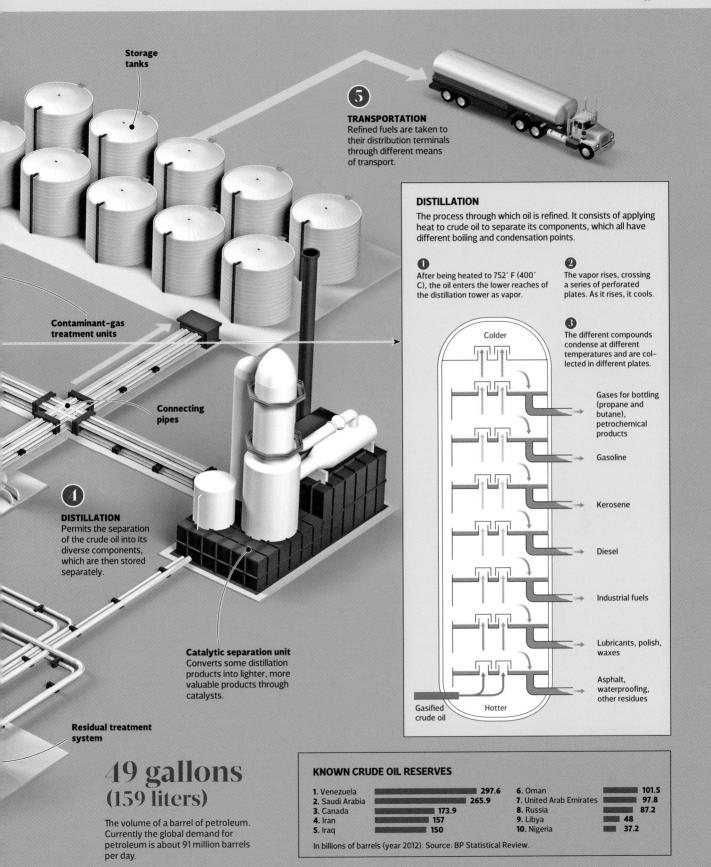

Storage tanks

⑤ TRANSPORTATION
Refined fuels are taken to their distribution terminals through different means of transport.

Contaminant-gas treatment units

Connecting pipes

① DISTILLATION
Permits the separation of the crude oil into its diverse components, which are then stored separately.

Catalytic separation unit
Converts some distillation products into lighter, more valuable products through catalysts.

Residual treatment system

DISTILLATION

The process through which oil is refined. It consists of applying heat to crude oil to separate its components, which all have different boiling and condensation points.

① After being heated to 752° F (400° C), the oil enters the lower reaches of the distillation tower as vapor.

② The vapor rises, crossing a series of perforated plates. As it rises, it cools.

③ The different compounds condense at different temperatures and are collected in different plates.

Colder

Gases for bottling (propane and butane), petrochemical products

Gasoline

Kerosene

Diesel

Industrial fuels

Lubricants, polish, waxes

Asphalt, waterproofing, other residues

Gasified crude oil

Hotter

49 gallons (159 liters)

The volume of a barrel of petroleum. Currently the global demand for petroleum is about 91 million barrels per day.

KNOWN CRUDE OIL RESERVES

1. Venezuela	297.6	
2. Saudi Arabia	265.9	
3. Canada	173.9	
4. Iran	157	
5. Iraq	150	
6. Oman	101.5	
7. United Arab Emirates	97.8	
8. Russia	87.2	
9. Libya	48	
10. Nigeria	37.2	

In billions of barrels (year 2012). Source: BP Statistical Review.

Nuclear reactor

One of the most efficient and cleanest methods for obtaining electric energy is through a controlled nuclear reaction. Although this technology has been used for half a century, it continues to be at the center of debate because of the risks it poses to the environment and health and because of the highly toxic waste it creates. Nuclear power generates a huge amount of radioactivity that can cause irreparable damages if released.

↓ MODERATOR

To break down the nucleus, the neutrons must collide with it at a specific speed, which is governed by a moderating substance, such as water, heavy water, graphite, and so on.

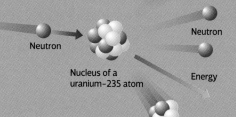

Neutron

Neutron

Nucleus of a uranium-235 atom

Energy

Fission

The nuclei of certain atoms, like uranium-235, can be broken apart when bombarded by neutrons. In doing so, they release great amounts of energy and new neutrons that can break down the nuclei of other atoms, creating a chain reaction. →

GENERATION OF ENERGY

The purpose of nuclear fission is to create very hot steam to operate turbines and electrical generators. The high temperatures are achieved by using nuclear energy from the reactor.

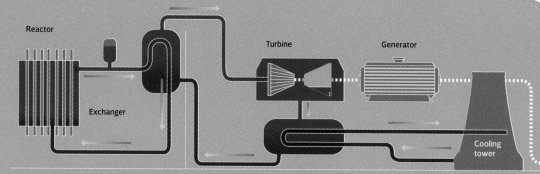

Reactor

Turbine

Generator

Exchanger

Cooling tower

 WATER
Pressurized water, together with the moderator, is pumped through the core of the reactor, and the temperature of the core increases by hundreds of degrees.

 STEAM
The resulting steam enters an exchanger, where it heats water until it too is converted into steam.

 ELECTRICITY
The steam enters the turbines, making them spin. The turbines drive the generator that produces electricity.

 RECYCLING
The steam condenses into liquid water and is reused.

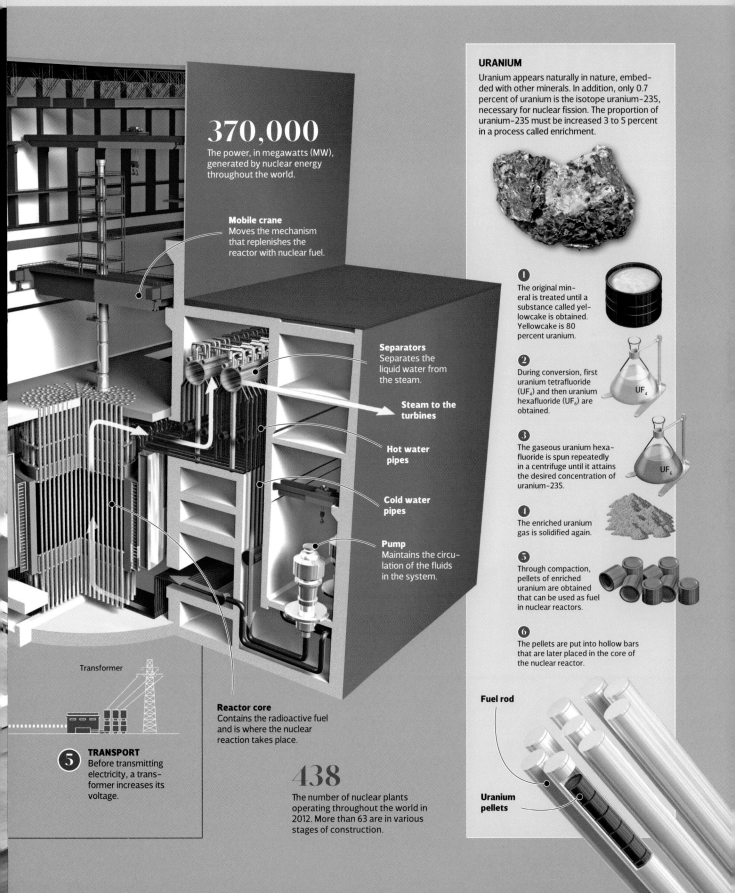

URANIUM

Uranium appears naturally in nature, embedded with other minerals. In addition, only 0.7 percent of uranium is the isotope uranium-235, necessary for nuclear fission. The proportion of uranium-235 must be increased 3 to 5 percent in a process called enrichment.

1 The original mineral is treated until a substance called yellowcake is obtained. Yellowcake is 80 percent uranium.

2 During conversion, first uranium tetrafluoride (UF_4) and then uranium hexafluoride (UF_6) are obtained.

3 The gaseous uranium hexafluoride is spun repeatedly in a centrifuge until it attains the desired concentration of uranium-235.

4 The enriched uranium gas is solidified again.

5 Through compaction, pellets of enriched uranium are obtained that can be used as fuel in nuclear reactors.

6 The pellets are put into hollow bars that are later placed in the core of the nuclear reactor.

Fuel rod

Uranium pellets

370,000
The power, in megawatts (MW), generated by nuclear energy throughout the world.

Mobile crane
Moves the mechanism that replenishes the reactor with nuclear fuel.

Separators
Separates the liquid water from the steam.

Steam to the turbines

Hot water pipes

Cold water pipes

Pump
Maintains the circulation of the fluids in the system.

Transformer

5 TRANSPORT
Before transmitting electricity, a transformer increases its voltage.

Reactor core
Contains the radioactive fuel and is where the nuclear reaction takes place.

438
The number of nuclear plants operating throughout the world in 2012. More than 63 are in various stages of construction.

Solar heating

Harnessing the energy of the sun to produce electricity and heat for everyday use on Earth is gaining in popularity. Applications for this clean, limitless form of energy range from charging batteries in telecommunications satellites to public transportation, all the way to building solar households in greater numbers throughout the world.

Energy regulator

Solar heat energy

Solar light is also used as a source to heat homes and the water inside them. In this case, solar collectors are used, which capture the heat from the sun, but don't produce electrical energy. To produce electricity, it is necessary to have solar or photovoltaic cells. →

SOLAR CELL

It is essentially formed by a thin layer of semiconductor material (silicon, for example), where the photovoltaic effect—the transformation of light into electrical energy—takes place.

① The sun shines on the cell. Some very energetic photons move the electrons and make them jump to the illuminated face of the cell.

② The negatively charged electrons generate a negative terminal on the illuminated face and leave an empty space in the positively charged dark face (positive terminal).

③ Once the circuit is closed, there is a constant flow of electrons (electric current) from the negative terminal to the positive one.

④ The current is maintained as long as the sun illuminates the cell.

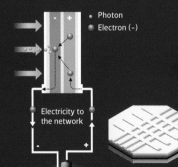

- Photon
- Electron (–)

Electricity to the network

Negative contact (–)

Active charge carrier zone

Upper metallic grid contact (negative electrode)

Upper metallic grid contact (positive electrode)

Positive contact (+)

Negative semiconductor (–) (mostly silicon)

Positive semiconductor (+) (mostly silicon)

Investment

One of the main problems with using solar energy on an industrial scale is the high startup cost required to harness the energy; this cost keeps solar energy from competing with other cheaper energy sources.

COLLECTOR

It works with the greenhouse effect. It absorbs the heat from the Sun and avoids losing that heat. While doing so, it heats a pipe where the fluid travels (water or gas), which heats a boiler (interchanger).

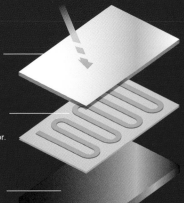

Protective shield
Formed by one or more glass plates, it allows the sunlight to go in and keeps the heat.

Absortion plate
It has piping, generally made of copper, where the fluid circulates, and warms up in the collector.

Thermic plate
The refracting material and the black color absorb the maximum heat of the sun. The protective shield avoids having losses as a result of that.

HOT WATER AND HEATING CIRCUIT

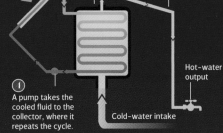

① The hot liquid flows from the collector through a circuit.

② It enters a heat exchanger, where it heats the water used in the house.

③ The water leaves the exchanger at a temperature suitable for domestic use or for heating a house.

④ A pump takes the cooled fluid to the collector, where it repeats the cycle.

Hot-water output

Cold-water intake

180° F
(82° C)

The maximum temperature a solar collector can reach when used to heat a house or to simply boil water.

← OTHER APPLICATIONS
In almost every system powered by electricity, solar energy can play a central role without endangering the environment. Although this technology is presently more expensive to use than coal, natural gas, or petroleum, this difference in cost could change soon.

Wind turbines

One of the most promising renewable energy resources is the use of wind to produce electricity by driving enormous wind turbines (windmills). Eolic power is an inexhaustible, clean, nonpolluting source of energy with more advantages than disadvantages. The most important disadvantages are our inability to predict precisely the force and direction of winds and the possibly negative scenic impact that groups of large towers could have on the local landscape.

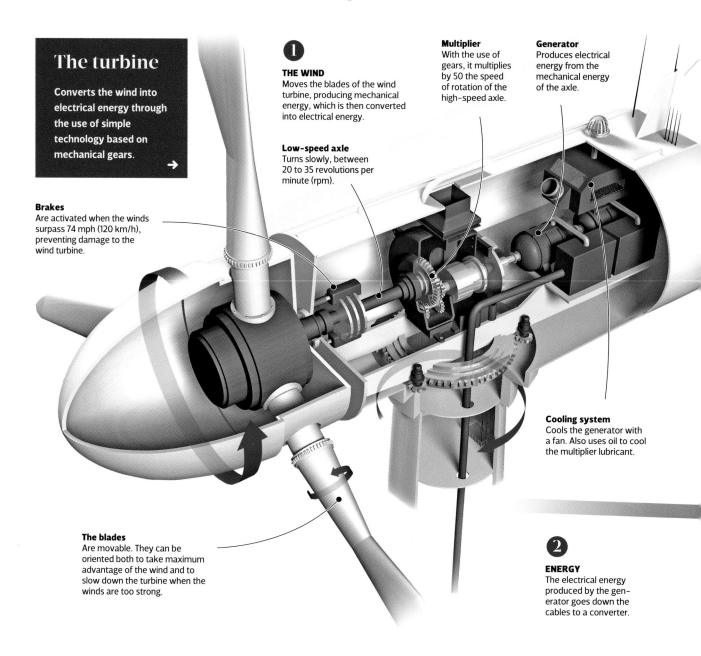

The turbine

Converts the wind into electrical energy through the use of simple technology based on mechanical gears. →

① THE WIND
Moves the blades of the wind turbine, producing mechanical energy, which is then converted into electrical energy.

Low-speed axle
Turns slowly, between 20 to 35 revolutions per minute (rpm).

Multiplier
With the use of gears, it multiplies by 50 the speed of rotation of the high-speed axle.

Generator
Produces electrical energy from the mechanical energy of the axle.

Brakes
Are activated when the winds surpass 74 mph (120 km/h), preventing damage to the wind turbine.

Cooling system
Cools the generator with a fan. Also uses oil to cool the multiplier lubricant.

The blades
Are movable. They can be oriented both to take maximum advantage of the wind and to slow down the turbine when the winds are too strong.

② ENERGY
The electrical energy produced by the generator goes down the cables to a converter.

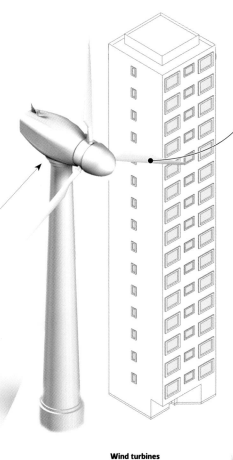

Blades
Measure, on average, 130 ft (40 m) in length. Three-blade rotors have proven to be the most efficient design.

370,000 Megawatts

Is the installed capacity of wind farms in the world. The leading country is China, followed by the United States and Germany.

WIND TURBINES

These modern, large wind turbines, between 150 and 200 ft (45 and 60 m) high, tend to be grouped in windy, isolated, mostly deserted regions. The most modern wind turbines can generate between 500 to 2,000 kW of power.

High terrain that is free of obstacles is ideal for wind turbines, because the wind blows freely there and reaches the turbines without turbulence.

The wind turbines are grouped into wind farms to maximize the potential of transmitting energy from only one location. This has the advantage of lowering costs and reducing environmental impact on the landscape.

↓ THE JOURNEY OF ELECTRICITY The energy produced in wind farms can travel through the main power grid together with energy generated by other sources.

Substations
Receive the energy from the collection plant and increase the voltage by hundreds of thousands of times for transmission to distant cities.

Wind turbines

The transformer
Increases by several thousand volts the voltage from the turbines.

Nearby cities
Receive the energy directly from the collection plant.

The collection plant
Receives the energy from all the transformers.

❸ GRID
After leaving the wind farm, the electrical energy can be incorporated into the main distribution grid.

❹ HOMES
The electricity reaches the residential distribution grid and finally homes.

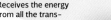

Hydroelectric energy

About 20 percent of the world's electricity is generated by the force of rivers through the use of hydroelectric power plants. This technology, used since the 19th century, employs a renewable, non-polluting resource, although the impact of dams on the environment is high. According to the United Nations, just two thirds of the world's hydroelectric potential is being used, most commonly in North America and Europe.

Diversion of the river

Charging chamber

Pipes

Powerhouse

River

Turbine room

The place where the kinetic energy of a river is transformed into mechanical energy by turbines and later into electrical energy by generators. →

Needle
Controls the pressure of the water injected into the wheels.

Wheel
The force of the water on the blades makes it spin.

① WATER
Enters the powerhouse under pressure and is injected into the turbine.

Injectors
Inject water under pressure onto the turbine wheel.

② TURBINE
The force of the water on its blades causes the turbine to turn.

③ ENERGY
The turbine makes the generator turn, thereby producing electric energy. The water is then returned to the river.

FROM THE DAM TO THE CITY

Electricity generated by the power plant is sent to a transformer, where its voltage is increased for transmission.

The electrical energy circulates through high-voltage power grids over great distances.

A transformer lowers the voltage of the electricity before distributing it to homes.

↓ BYPASS PLANT

Does not have a reservoir. It simply takes advantage of the available flow of water and is thus at the mercy of seasonal variations in water flow. It also cannot take advantage of occasional surplus water.

↓ PLANTS WITH RESERVOIRS

The presence of a reservoir, formed by a containment dam, guarantees a constant flow of water—and therefore energy—independent of variations in water level.

China

The world's largest producer of hydroelectricity (95,000 MW installed), followed by the United States, Canada, and Brazil.

①
The water enters the powerhouse and turns the turbines. The generators produce electricity.

②
Once used, the water is returned to the river.

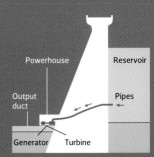

Powerhouse
Reservoir
Output duct
Pipes
Generator
Turbine

Reservoir

Dam

Pipes

Powerhouse

22,500

The hydroelectric capacity, in megawatts, of China's Three Gorges Dam, which was completed in 2012. Previous record holder was the 12,600-MW Itaipú Dam on the border between Paraguay and Brazil.

↓ PUMPING PLANT

Has two reservoirs located at different levels. In this way, the water can be reused, which allows a more efficient management of water resources.

Generator
Transforms the mechanical energy of the turbines into electrical energy.

①
The water goes from the upper reservoir to the lower one, generating electricity in the process.

Powerhouse
Reservoir
Second reservoir
Pipes
Turbine

Reservoir

Dam

Pipes

Powerhouse

Second reservoir

②
In off-peak hours, the water is pumped to the first reservoir to be reused.

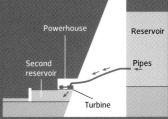

Powerhouse
Reservoir
Second reservoir
Pipes
Turbine

Geothermal energy

Geothermal energy is one of the cleanest and most promising sources of energy. The first geothermal plant started operating more than 100 years ago. Geothermal plants generate electricity from the heat that emanates from the Earth's interior. Geothermal power plants, however, suffer from some limitations, such as the fact that they must be constructed in regions with high volcanic activity. The possibility of geothermal power plants becoming useless due to a reduction in such volcanic activity is always present.

TYPES OF POWER PLANTS

Not all geothermal power plants are identical. Their design depends on the type of geothermal deposit from which the energy is extracted.

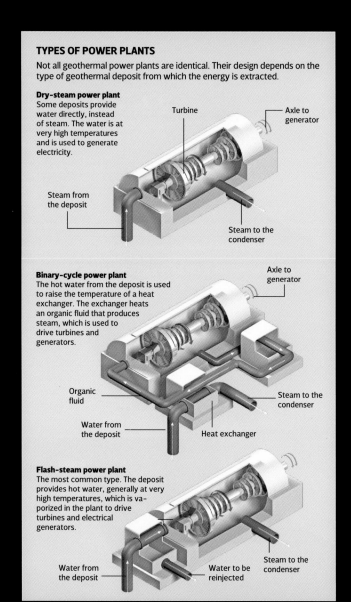

Dry-steam power plant
Some deposits provide water directly, instead of steam. The water is at very high temperatures and is used to generate electricity.

Turbine

Axle to generator

Steam from the deposit

Steam to the condenser

Binary-cycle power plant
The hot water from the deposit is used to raise the temperature of a heat exchanger. The exchanger heats an organic fluid that produces steam, which is used to drive turbines and generators.

Axle to generator

Organic fluid

Steam to the condenser

Water from the deposit

Heat exchanger

Flash-steam power plant
The most common type. The deposit provides hot water, generally at very high temperatures, which is vaporized in the plant to drive turbines and electrical generators.

Water from the deposit

Water to be reinjected

Steam to the condenser

Deposits

Accumulated underground water and steam, sometimes contained in cracks or porous rocks, is heated by the magma and can be used as a renewable energy resource.

→

↓ TYPES OF GEOTHERMAL DEPOSITS

Geothermal deposits are classified by their temperature and by the resource they provide (water or steam).

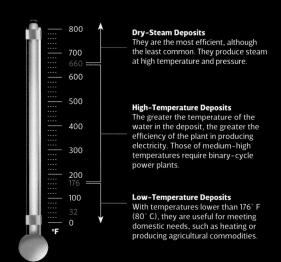

800
700
660
600
500
400
300
200
176
100
32
0
°F

Dry-Steam Deposits
They are the most efficient, although the least common. They produce steam at high temperature and pressure.

High-Temperature Deposits
The greater the temperature of the water in the deposit, the greater the efficiency of the plant in producing electricity. Those of medium-high temperatures require binary-cycle power plants.

Low-Temperature Deposits
With temperatures lower than 176° F (80° C), they are useful for meeting domestic needs, such as heating or producing agricultural commodities.

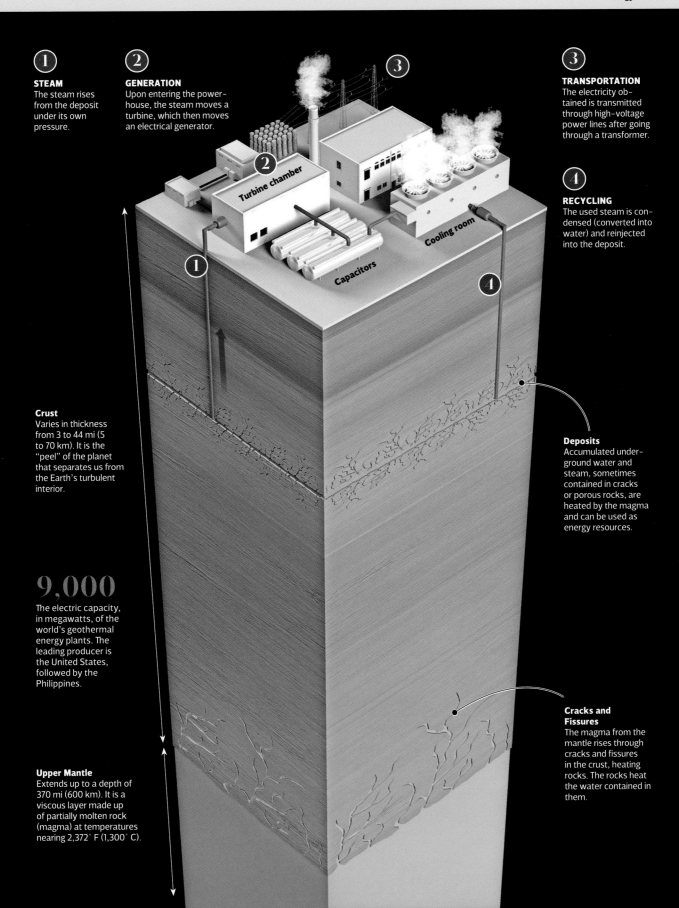

① STEAM
The steam rises from the deposit under its own pressure.

② GENERATION
Upon entering the power-house, the steam moves a turbine, which then moves an electrical generator.

③ TRANSPORTATION
The electricity obtained is transmitted through high-voltage power lines after going through a transformer.

④ RECYCLING
The used steam is condensed (converted into water) and reinjected into the deposit.

Turbine chamber

Capacitors

Cooling room

Crust
Varies in thickness from 3 to 44 mi (5 to 70 km). It is the "peel" of the planet that separates us from the Earth's turbulent interior.

Deposits
Accumulated underground water and steam, sometimes contained in cracks or porous rocks, are heated by the magma and can be used as energy resources.

9,000
The electric capacity, in megawatts, of the world's geothermal energy plants. The leading producer is the United States, followed by the Philippines.

Cracks and Fissures
The magma from the mantle rises through cracks and fissures in the crust, heating rocks. The rocks heat the water contained in them.

Upper Mantle
Extends up to a depth of 370 mi (600 km). It is a viscous layer made up of partially molten rock (magma) at temperatures nearing 2,372° F (1,300° C).

Tidal power plant

The variations in the tides and the force of the oceans' waves contain an enormous amount of potential energy for generating electricity. Tidal power neither emits pollution into the atmosphere, nor depletes resources, as happens in with fossil fuels. Tidal plants are similar to hydroelectric plants. They have a water-retention dam (which crosses an estuary from shore to shore) and a powerhouse where the turbines and generators, which produce electricity, are located.

Location of the Dam

The power plant needs be located in a river outlet to the sea (estuary) or in a narrow bay—places that have above-average tidal amplitude (the variance between low tide and high tide).

Gates
Are opened to let the water in as the tide rises and then closed to prevent its exit.

Tidal Power Plant
The turbines, which power the generators, are found inside the plant. They convert the kinetic energy of the water into mechanical energy and then into electrical energy.

12 hours 25 minutes

The approximate time between two high tides or two low tides, depending on the geographic location and sometimes on other factors, such as winds and ocean currents.

Foundations
Are built from concrete to prevent the erosion produced by the flow of water over the terrain.

Gates
Regulate the exit of trapped water through the turbines during the generation of electricity.

Turbines
Are powered by the flow of the water. When they turn, they move the generators that produce electricity.

Sihwa Lake

The largest tidal power plant in the world opened in 2011. It is located in South Korea and has an electrical generating capacity of 254 megawatts.

← AMPLITUDE OF THE TIDES To produce electricity efficiently, the variance between high and low tide needs to be at least 13 feet (4 m); this variation limits the number of possible locations for tidal plants.

Electrical Substation
Increases the voltage of the generated power before its transmission.

High-Voltage Grid
Takes the electrical energy to the regions where it will be consumed.

Dam
Crosses the estuary or bay from shore to shore. It retains the water during high tide.

GENERATION OF ELECTRICITY

As in a hydroelectric power plant, the trapped water turns a turbine that operates the generators.

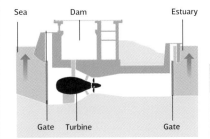
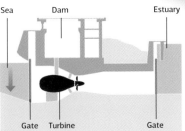
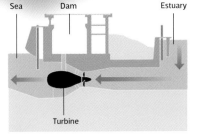

Sea Dam Estuary
Gate Turbine Gate

HIGH TIDE
During high tide, the level of the water rises in the estuary. The gates of the dam are opened to let the water in.

WATER RESERVOIR
Once high tide is over, the water level in the estuary begins to drop. The gates of the dam are closed to prevent the trapped water from escaping.

GENERATION
During low tide, the trapped water is released and it passes through the system of turbines that power the electrical generators.

Biodigesters

When anaerobic bacteria (bacteria that do not require oxygen to live) decompose organic material, through processes such as rotting and fermentation, they release biogas that can be used as an energy resource for heating and for generating electricity. They also create mud with very high nutritional value, which can be used in agriculture or fish farming. This technology appears promising as an energy alternative for rural and isolated regions, where, in addition to serving the energy needs of the populace, it helps recycle organic wastes.

Exeter

In 1895, this English city was the first to inaugurate a public lighting system powered by biogas (from a water-purification plant).

Dome
Is built underground and can be lined with concrete, brick, or stone.

The reactor

Is a closed chamber where the bacteria breaks down the waste. The generated gas (called biogas) and the fertilizing mud are collected for later use. ↓

WASTE
The organic waste is introduced into the reactor and mixed with water.

DIGESTION CHAMBER
Where the bacteria ferment the waste. They produce gas and fertilizing mud.

BIOGAS
Is a product of the process that contains methane and carbon dioxide. It is used for cooking, heating, and generating electricity.

FERTILIZING MUD
Very rich in nutrients and odorless, it is ideal for agricultural uses.

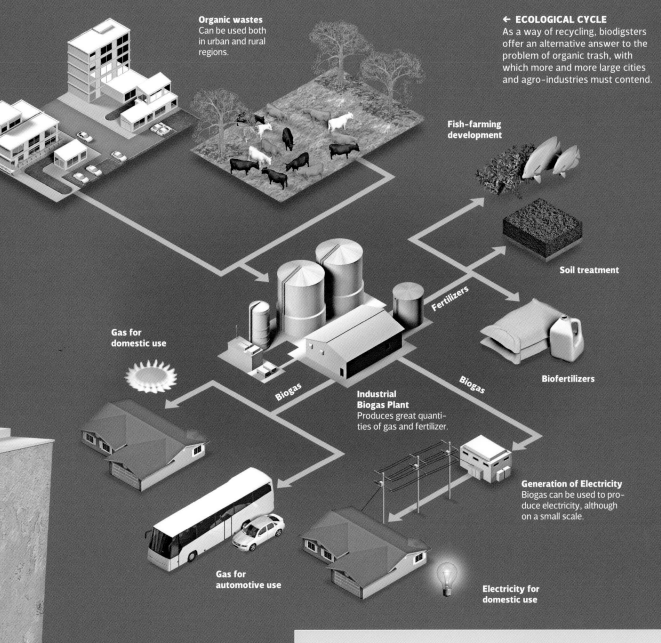

Organic wastes
Can be used both in urban and rural regions.

← ECOLOGICAL CYCLE
As a way of recycling, biodigsters offer an alternative answer to the problem of organic trash, with which more and more large cities and agro-industries must contend.

Fish-farming development

Soil treatment

Fertilizers

Gas for domestic use

Biofertilizers

Biogas

Biogas

Industrial Biogas Plant
Produces great quantities of gas and fertilizer.

Generation of Electricity
Biogas can be used to produce electricity, although on a small scale.

Gas for automotive use

Electricity for domestic use

Pathogens

Laboratory tests demonstrated that the biodigestion process kills up to 85 percent of the harmful pathogenic agents present in organic waste, pathogens which would otherwise be released into the environment.

← BIOGAS The gaseous product of biodigestion, it is made up of a mixture of gases whose makeup depends on the composition of the wastes and the break-down process: 55-70% Methane (CH_4), 30-45% Carbon Dioxide (CO_2), 1-10% Hydrogen (H_2), 0.5-3% Nitrogen (N_2), 0.1% Sulfuric Acid (H_2S).

FOOD INDUSTRY

Fishing

The international demand for fish and shellfish has encouraged the use of highly efficient fishing vessels and techniques. The use of these vessels and techniques, however, has caused ever-increasing destruction of the environment and fish breeding grounds. Every year, fishing nets kill more than 300,000 whales, dolphins, and porpoises worldwide. The greatest threat facing many water species is to become enmeshed in nets meant to catch fish for human consumption.

↑ OVERFISHING The fishing industry is an important source of food and employment around the world, and it provides the world's population with 16 percent of all animal protein consumed. However, environmental pollution, climate change, and irresponsible fishing practices are taking their toll on the planet's marine resources.

10%
Of all fish species are extinct or recovering.

Local vessels
Catch fish in surface waters. The fish they catch are usually sold in the surrounding area.

Stone traps
Strand schools of small fish when the tide goes out.

Raking cockles
Cockles and other shellfish can be gathered at low tide by raking the sand.

Traditional fishing

Traditional fishing is a widespread, small-scale activity practiced directly by fishermen using selective fishing techniques. Such harvesting of fish and shellfish is carried out with equipment such as harpoons, hand nets, fishing rods, and fish traps. The vessels may include anything from pirogues to small motorboats. ↗

Net traps
Are a series of cone-shaped nets with a cylinder at one end. They trap fish that swim with the current.

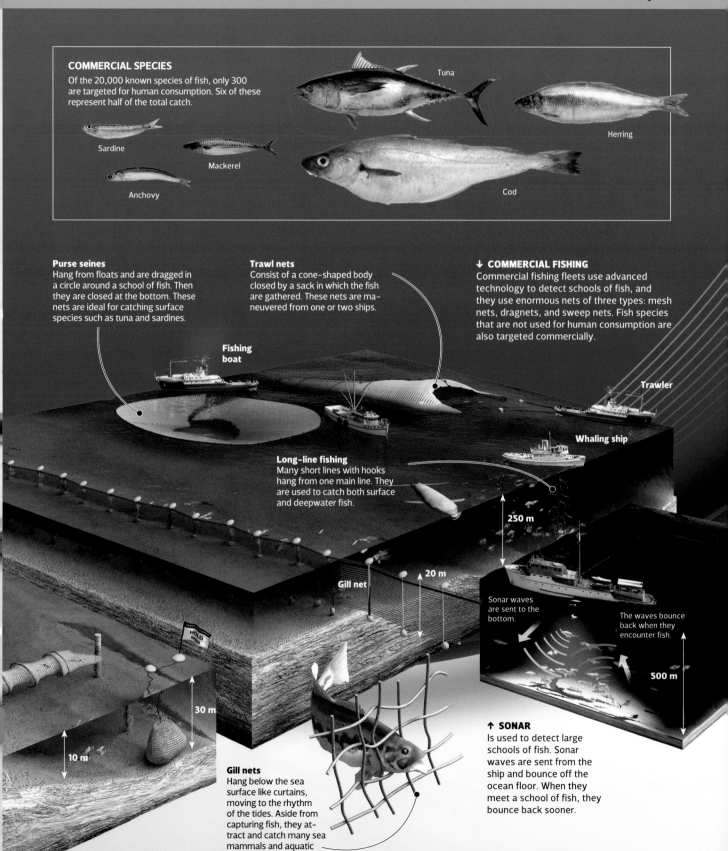

COMMERCIAL SPECIES

Of the 20,000 known species of fish, only 300 are targeted for human consumption. Six of these represent half of the total catch.

Tuna

Herring

Sardine

Mackerel

Anchovy

Cod

Purse seines
Hang from floats and are dragged in a circle around a school of fish. Then they are closed at the bottom. These nets are ideal for catching surface species such as tuna and sardines.

Trawl nets
Consist of a cone-shaped body closed by a sack in which the fish are gathered. These nets are maneuvered from one or two ships.

↓ COMMERCIAL FISHING
Commercial fishing fleets use advanced technology to detect schools of fish, and they use enormous nets of three types: mesh nets, dragnets, and sweep nets. Fish species that are not used for human consumption are also targeted commercially.

Fishing boat

Trawler

Whaling ship

Long-line fishing
Many short lines with hooks hang from one main line. They are used to catch both surface and deepwater fish.

250 m

20 m

Gill net

Sonar waves are sent to the bottom.

The waves bounce back when they encounter fish.

500 m

30 m

10 m

Gill nets
Hang below the sea surface like curtains, moving to the rhythm of the tides. Aside from capturing fish, they attract and catch many sea mammals and aquatic birds, which then die.

↑ SONAR
Is used to detect large schools of fish. Sonar waves are sent from the ship and bounce off the ocean floor. When they meet a school of fish, they bounce back sooner.

Tomato factories

The colonization of the Americas brought about the discovery of an extraordinary variety of plants that have been used as food ever since. An important example is the tomato, which is consumed globally. The cultivation of the tomato has reached marked levels of technological complexity that help address problems of infestation and adverse environmental conditions, as well as make it possible to grow tomatoes without using soil.

Traditional cultivation

In gardens, tomato plants are grown in accordance with their annual growth cycle, using adequate soil and pest control.

Greenhouse
Seedlings grow, protected from frosts.

Fertilizer
Provides the soil with nutrients.

PLANTING
End of Winter

HARVESTING
Beginning of Summer

Irrigation
Each plant requires more than 0.5 g (2 l) of water every week as it grows.

Good neighbors
Raising carrot and cabbage crops in the same garden aids the development of tomatoes.

Transplant
The seedling can be transplanted when it has three or four real leaves.

Stakes
Help the plants to grow and remain upright.

Level A
Has nutrients that are essential to the plant.

Level B
Allows for good water drainage from rain or irrigation.

Sandy loam soil
Allows for the best development of tomatoes.

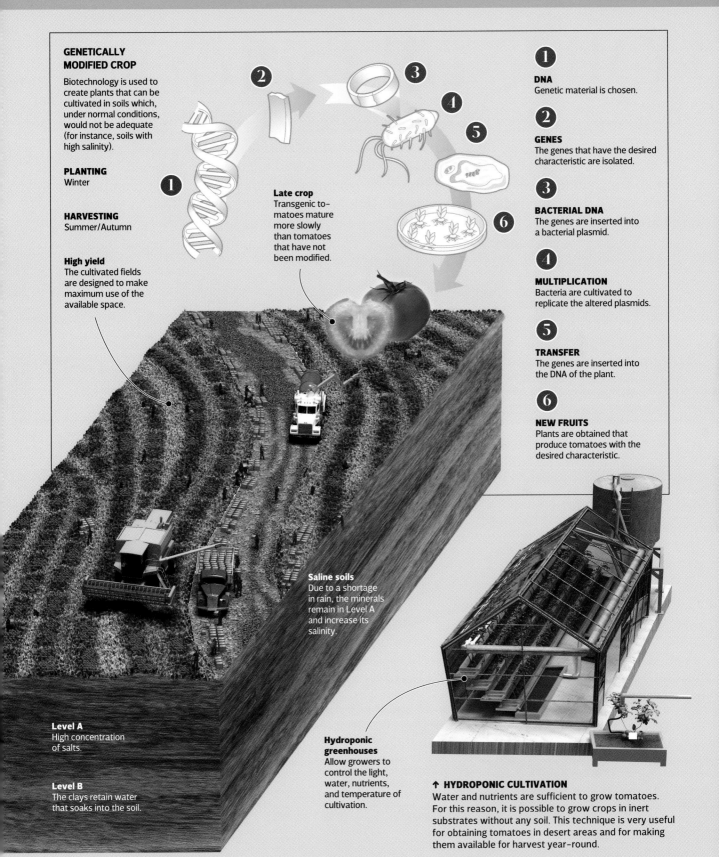

GENETICALLY MODIFIED CROP

Biotechnology is used to create plants that can be cultivated in soils which, under normal conditions, would not be adequate (for instance, soils with high salinity).

PLANTING
Winter

HARVESTING
Summer/Autumn

High yield
The cultivated fields are designed to make maximum use of the available space.

Late crop
Transgenic tomatoes mature more slowly than tomatoes that have not been modified.

1 DNA
Genetic material is chosen.

2 GENES
The genes that have the desired characteristic are isolated.

3 BACTERIAL DNA
The genes are inserted into a bacterial plasmid.

4 MULTIPLICATION
Bacteria are cultivated to replicate the altered plasmids.

5 TRANSFER
The genes are inserted into the DNA of the plant.

6 NEW FRUITS
Plants are obtained that produce tomatoes with the desired characteristic.

Saline soils
Due to a shortage in rain, the minerals remain in Level A and increase its salinity.

Level A
High concentration of salts.

Level B
The clays retain water that soaks into the soil.

Hydroponic greenhouses
Allow growers to control the light, water, nutrients, and temperature of cultivation.

↑ HYDROPONIC CULTIVATION

Water and nutrients are sufficient to grow tomatoes. For this reason, it is possible to grow crops in inert substrates without any soil. This technique is very useful for obtaining tomatoes in desert areas and for making them available for harvest year-round.

Olive oil factories

Olive oil has been a part of people's diet since antiquity, and today it remains one of the most popular oils because of its flavor and nutritious properties. Obtaining high-quality olive oil involves a chain of processes that begins at the tree and ends with the packaging of the end product. The quality of the olive oil is determined by several factors, starting in the fields and depends on a combination of soil, climate, oil variety, and cultivation and harvesting techniques. The remaining operations in the extraction process (transportation, storage, manufacturing, and extraction of the oil) are responsible for maintaining that quality.

Olive fruit

Olive oil, known for its quality, is the main product of the olive fruit. Approximately 22 lbs (10 kg) of olives produces 2.11 qt (2 l) of oil. ↘

① CULTIVATION
Plowed land, a moderate climate, an altitude of up to 2,300 feet (700 m) above sea level, and up to 15 inches (40 cm) of rain per year sum up the conditions needed for the development of olive trees.

② WASHING AND CLASSIFICATION
The fruits are carefully washed with water and then classified according to their variety.

③ MILLING
Machines break open the fruit and mix it to create a homogenous paste. This must be done on the day the fruit is harvested.

Stone wheel
Hammer systems are also used.

Collection
Harvesting is done by hitting the tree branches, either by hand or mechanically, so that the fruits fall to the ground.

New plantings
Are propagated through staking, layering, or the taking of cuttings.

← THE QUALITY OF THE OIL The oil that comes out of the first pressing from good quality fruits and with an acid level lower than 0.8 percent is called extra virgin. After this pressing the other levels of oil quality are obtained.

95 %
Of the world's product comes from the Mediterranean. Spain, Italy, and Greece are the world's main producers.

OLIVE GROWTH STAGES (IN THE SOUTHERN HEMISPHERE)

1
Flowers
Are distributed in clusters of 10 to 40.

May

2
Growth
The pit or drupe (endocardium) has hardened; the fruit grows.

July

August

September

3
Green olive
The appearance of this color tells us the fruit is edible.

October

4
Maturing
Purple spots begin to show.

November

5
Mature fruit
The oxidation process has given it a black color.

Seed

December

→ PRESS
This press has a hydraulic mechanism that compresses the disks.

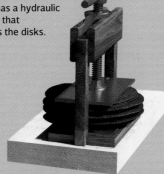

4
PRESSING
Traditionally, the paste that includes the entire olive is placed on a system of stacked disks and then compressed by a hydraulic press.

5
REFINING
The oil obtained is separated from the other solid residues, impurities, and water. Since antiquity, this process has been carried out by decantation, which requires letting the oil sit undisturbed after it comes out of the press. Today it can also be carried out with vertical centrifuges.

Filter
Centrifuges are now used.

Residue
Can be used to obtain other oils.

Homogenizing
The oil from several hoppers is mixed in the final stage to obtain a uniform product.

6
STORAGE
Virgin olive oil has nonfat components that have to be preserved during storage and packaging. It must be kept in a dark place at a stable temperature.

4

5

6

7

Bottle
This is how the oil is sent to the market.

Stainless steel hopper
The residues are decanted at a temperature that is low, but not too low: oil crystallizes between 32° and 36° F (0° and 2° C).

7
BOTTLING
Is carried out in a plant, although sometimes it is done manually to ensure product quality. Glass, aluminum, and plastic containers are used. Bottles cannot be stored where they will be exposed to light, odors, or heat for extended periods.

New agriculture

With the invention of agriculture some 10,000 years ago, people began to grow their own food. They learned that if they developed the proper technology, they could increase the productivity of cropland and thereby make more food available. They also learned that the more intensive their farming practices were, the faster the land would be worn out and the more rapidly it would lose its fertility. New agricultural technology can overcome both of these problems, although there is, as yet, no method capable of solving all of the problems that agriculture faces.

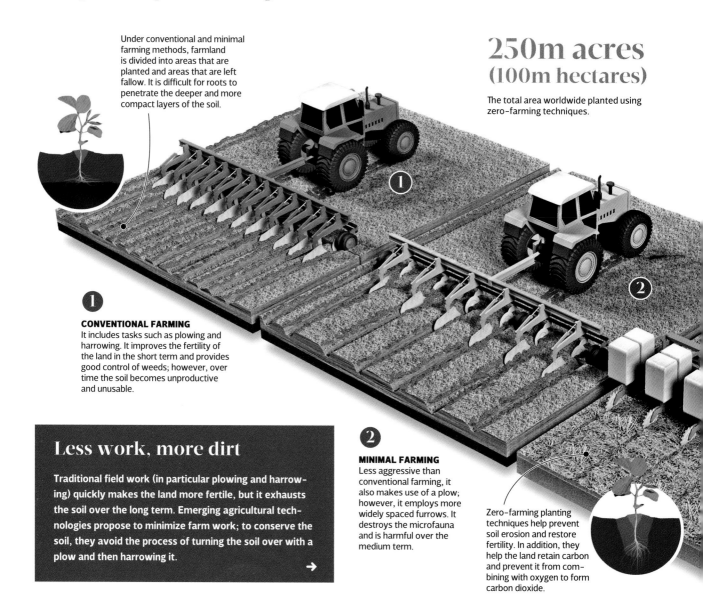

Under conventional and minimal farming methods, farmland is divided into areas that are planted and areas that are left fallow. It is difficult for roots to penetrate the deeper and more compact layers of the soil.

250m acres
(100m hectares)

The total area worldwide planted using zero-farming techniques.

1 CONVENTIONAL FARMING
It includes tasks such as plowing and harrowing. It improves the fertility of the land in the short term and provides good control of weeds; however, over time the soil becomes unproductive and unusable.

Less work, more dirt

Traditional field work (in particular plowing and harrowing) quickly makes the land more fertile, but it exhausts the soil over the long term. Emerging agricultural technologies propose to minimize farm work; to conserve the soil, they avoid the process of turning the soil over with a plow and then harrowing it. →

2 MINIMAL FARMING
Less aggressive than conventional farming, it also makes use of a plow; however, it employs more widely spaced furrows. It destroys the microfauna and is harmful over the medium term.

Zero-farming planting techniques help prevent soil erosion and restore fertility. In addition, they help the land retain carbon and prevent it from combining with oxygen to form carbon dioxide.

GENETICALLY MODIFIED CROPS

Despite controversy and worldwide campaigns against them, genetically modified crops have been a boon in countries such as the United States, Brazil, and Argentina. These genetically treated varieties acquire new qualities that make them more efficient to market and sell.

How they works

Genetically modified crops have distinctive qualities. Examples are long-lasting tomatoes or dwarf sunflowers that are not bothered by the wind. Another example is the genetically modified soybean, which is resistant to pesticides.

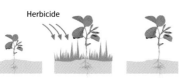

Herbicide

All the weeds die except the transgenic soybean, which contains a gene that makes it resistant to the herbicide.

The major criticism of zero farming is that, as the land's fertility improves, there is a related increase in weeds and pests and thus a greater need to apply agrochemicals to control them.

High yield

Low yield

Planting

1 The disk makes a cut in the organic bed about 4 inches (10 cm) deep.

2 The dual disk deposits the seed in its furrow at a precise depth.

3 The toothed steel wheels close the furrow.

4 A dosimeter applies a small amount of pesticide and herbicide.

← PRECISION AGRICULTURE

It is possible to improve the productivity of farmland by using global positioning systems (GPS).

A harvester equipped with a GPS can make maps of crop yields. Plots with relatively low yields due to a lack of water or fertilizer can be revealed. The amount of fertilizer and water applied to low-yield plots can then be altered to improve the overall efficiency of the farm.

The illustration shows the yield of different areas of a cornfield. Using information provided by GPS, the farmer can make the necessary adjustments to maximize the yield throughout the cornfield.

↑ **ORGANIC FARMING** Another popular trend is organic farming—that is, farming that uses no fertilizers or synthetic pesticides. Organic farming employs natural strategies for fertilizing and pest control.

3

ZERO FARMING

The land is not plowed or cultivated with a harrow. Instead, it is left with the residue from the previous crop. Over time this material forms a moist, nourishing, organic bed that protects the soil from erosion. It does not destroy the microfauna and flora.

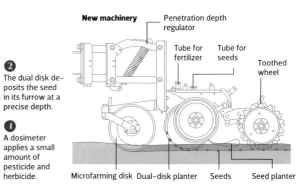

New machinery — Penetration depth regulator

Tube for fertilizer

Tube for seeds

Toothed wheel

Microfarming disk   Dual-disk planter   Seeds   Seed planter

The algae industry

In addition to other uses, algae are part of food and traditional medicines in many Asian countries. There are different types of algae, depending on the waters where they live. Algae can contain 25% minerals and trace elements and proteins (depending on the species). They also contain many amino acids, and a large supply of vitamins. The seaweed industry started in Japan in the seventeenth century, spreading a century later to the West for the extraction of iodine and other chemicals of high economic value.

Agar

Algae and their extracts are used to produce food, medicines, and cosmetics. Most algae is still collected by hand, although many of the large species are collected with special boats, such as that used to obtain agar, a gelatinous substance used as a thickener for vegetable gelatins, ice cream and desserts. ↘

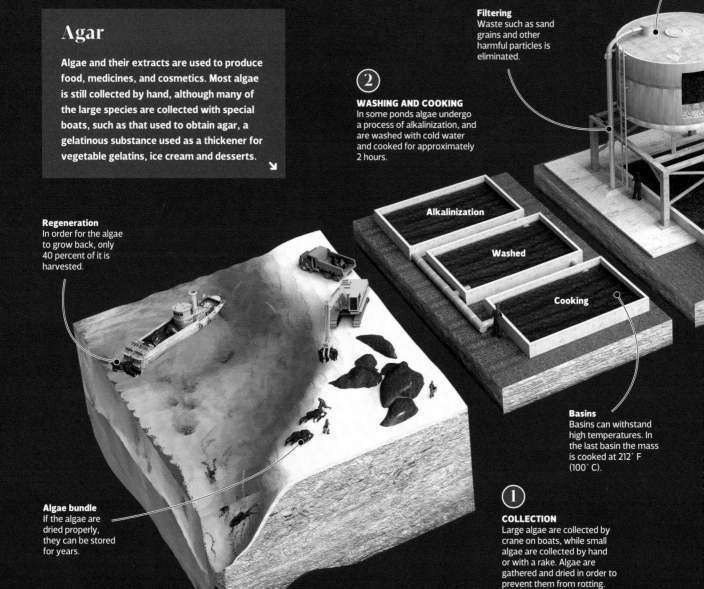

Filtering
Waste such as sand grains and other harmful particles is eliminated.

②

WASHING AND COOKING
In some ponds algae undergo a process of alkalinization, and are washed with cold water and cooked for approximately 2 hours.

Alkalinization

Washed

Cooking

Regeneration
In order for the algae to grow back, only 40 percent of it is harvested.

Algae bundle
If the algae are dried properly, they can be stored for years.

Basins
Basins can withstand high temperatures. In the last basin the mass is cooked at 212° F (100° C).

①

COLLECTION
Large algae are collected by crane on boats, while small algae are collected by hand or with a rake. Algae are gathered and dried in order to prevent them from rotting.

150 pounds

Per square inch (10 kg/sq cm) is the pressure at which hot air is applied to dry the mass.

Crushed algae
Bleaching with salt water improves its quality.

Grinding apparatus
Dry agar powder is passed through a grinder several times to reduce the grain size.

Quality control
Samples are taken during successive stages of sifting.

Pool
The pool receives the mixture free from rock or shell matter. A mechanism in the tank slowly stirs the mixture.

Gelling
Occurs when the temperature is lowered along the length of the pipe to 77° F (25° C).

Drying belt
Hot air 160–175° F (70–80° C).

Moist gel

Press

Drying press

Gelatin
Contains 1% agar.

④

DRYING
Gel sheets about 0.4 inch (1 cm) wide come out of the press between layers of nylon. They are placed on platforms, where they begin to dry. The sheets are then placed on a conveyor belt and further dried by a stream of hot air.

⑤

FINISHING
Ground into a powder, the product must go through successive milling and sifting steps to eliminate any lumps and impurities. Samples are taken as the algae product is refined. Once it has passed inspection, the final product is packaged.

Proteins

The protein content in algae varies between 4% and 70%, depending on the species.

③

TRANSFORMATION
The filtrate contains only water and the algae extract. After a series of processes when the solution has cooled, a gel is obtained containing 1% agar. The agar gel is then pressed into 1cm plates.

9 pounds
(4 kg)

The quantity of fresh algae needed to obtain about 2 pounds (1kg) of dry algae.

RICH VARIETY

Wakame
Dried in the sun, it is most commonly used in the kitchen to prepare salads and soups. It is the richest in vitamin B12 and calcium, and is suitable for children and pregnant women.

Konbu
Dried in the sun and cut into long strips, it is also used grated, fried and marinated in vinegar chips. It is a source of vitamin B12, and rich in alginic acid, magnesium and iodine.

Nori
A red algae that is dried and packaged in small squares. It is rich in amino acids, calcium, phosphorus, vitamins A, C, D, B1, B2 and proteins. It is used as wrapper for sushi and improves memory.

SCIENCE

Weather station

The study of weather and climate is known as meteorology. Most of the information available regarding climate data comes from the record that meteorologists everywhere in the world keep regarding cloud cover, temperature, the force and direction of the wind, air pressure, visibility, and precipitation. Each meteorological station broadcasts the data by radio or satellite, making it possible for meteorologists to make forecasts and weather maps.

Workplace

A typical weather station checks the temperature, humidity, wind velocity and direction, solar radiation, rain, and barometric pressure. In some places, soil temperature and flow of nearby rivers are also monitored. The compilation of this data makes it possible to predict different meteorological phenomena.

ANEROID BAROMETER
Measures atmospheric pressure. Changes are shown by the arrows.

Weather Station
Meteorologists collect data at different heights. They use various instruments at ground level: a thermometer for temperature, a hygrometer for humidity, and a barometer for atmospheric pressure.

Barograph
Measures the atmospheric pressure and records its changes over time.

The light strikes and is concentrated as it traverses the sphere.

← HELIOPHANOGRAPH
An instrument used to measure the number of hours of sunlight. It has a glass sphere that acts as a lens to concentrate sunlight. The light is projected onto a piece of cardboard behind the sphere. The cardboard is then burned according to the intensity of the light.

Atmometer
Also known as an evaporimeter, it measures the effective rate water evaporates in the open air, from its loss from due to its transformation to water vapor.

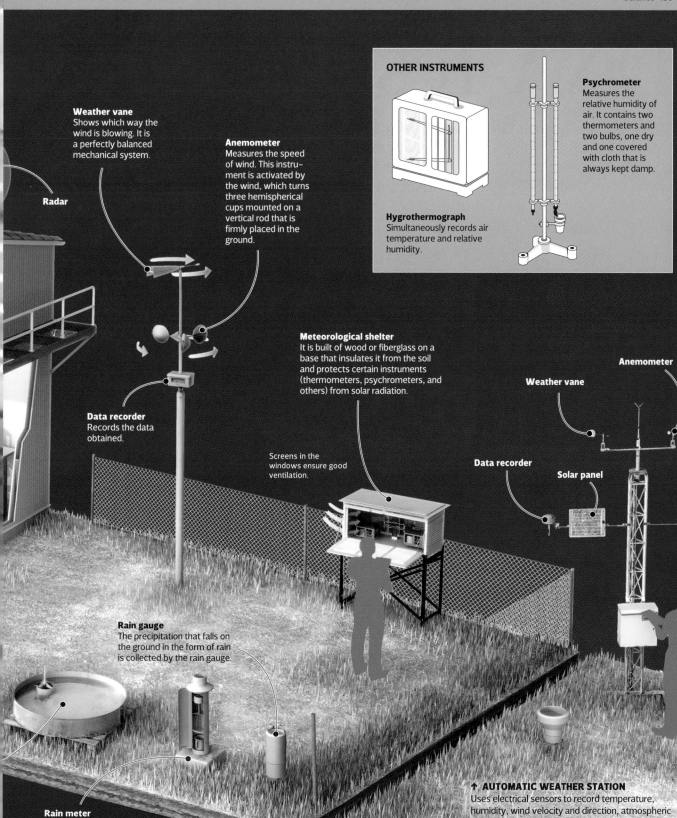

OTHER INSTRUMENTS

Psychrometer
Measures the relative humidity of air. It contains two thermometers and two bulbs, one dry and one covered with cloth that is always kept damp.

Hygrothermograph
Simultaneously records air temperature and relative humidity.

Weather vane
Shows which way the wind is blowing. It is a perfectly balanced mechanical system.

Anemometer
Measures the speed of wind. This instrument is activated by the wind, which turns three hemispherical cups mounted on a vertical rod that is firmly placed in the ground.

Radar

Meteorological shelter
It is built of wood or fiberglass on a base that insulates it from the soil and protects certain instruments (thermometers, psychrometers, and others) from solar radiation.

Anemometer

Weather vane

Data recorder
Records the data obtained.

Screens in the windows ensure good ventilation.

Data recorder

Solar panel

Rain gauge
The precipitation that falls on the ground in the form of rain is collected by the rain gauge.

Rain meter
This is used to keep a chronological record of the amount of water falling as rain.

↑ AUTOMATIC WEATHER STATION
Uses electrical sensors to record temperature, humidity, wind velocity and direction, atmospheric pressure, and rainfall, among other parameters. The readings are processed by microprocessors and transmitted via an automatic system 24 hours a day.

Cow cloning

The term "cloning" itself provokes controversy. Strictly speaking, to clone is to obtain an identical organism from another through technology. The most commonly used technique is called somatic-cell nuclear transfer. It was used to create Dolly, the sheep, as well as other cloned animals, including these Jersey cows. The technique consists of replacing the nucleus of an egg with the nucleus of a cell from a donor specimen. When the egg then undergoes division, it gives rise to an organism identical to the donor. With all such processes, there exist slight differences between the donor and the clone. In only one case is the clone perfect, and it comes naturally: monozygous (identical) twins.

①

OBTAINING THE NUCLEUS
A specialized cell of an adult animal, whose DNA is complete, is isolated, and it is cultivated in vitro to multiply it. Various eggs of a donor cow are also isolated. The nucleus is then removed from both groups of cells.

Nucleus of the cell to clone
The nucleus is transferred to the egg.

Nucleus extraction
A fibroblast is extracted from the ear of an exemplary adult.

Nucleus with complete DNA (60 chromosomes)

Egg without nucleus

Egg without nucleus
Only the cytoplasm, with organelles like mitochondria, remains.

Egg extraction
An egg is obtained from the ovary of another exemplary specimen, and the nucleus is removed.

②

NUCLEUS TRANSFER
Consists of replacing the nucleus of the egg with that obtained from the adult cell. In this form, the chromosomes carried by the new nucleus complete the egg in the same way as if it had been fertilized naturally. Once fused, the cell will begin its program of division as if it were a zygote (fertilized egg).

Cost

The technology is still inefficient. For this Jersey, 934 eggs were transferred, of which 166 fused, and only one was successfully developed.

Diverses uses

Cloning can be useful to obtain new organisms and tissues and for reproducing segments of DNA. ↓

Pipette
It is used to introduce the new nucleus into the egg.

16 CELLS

④ **CULTIVATION**
Cells are cultivated in vitro until the embryonic structure has formed and has the right size to be transplanted (the blastocyst). This process takes about one week.

8 CELLS

2 CELLS

③ **FUSION**
By means of light electric discharges, fusion of the donated nucleus with the cytoplasm of the egg is initiated. Three hours later, calcium is added to the cell to simulate fertilization. An interchange begins between the nucleus and the cytoplasm, and the cell starts to divide.

⑤ **INSEMINATION**
The blastocyst is implanted in the uterus of the mother donor. If all goes accordingly it attaches to uterine wall and continues its development.

⑥ **FETAL DEVELOPMENT**
Once it has been implanted, the blastocyst begins to grow. Because all the genetic information has been received from a single donor cell, the calf to be formed will be a genetically identical copy to the donor nucleus.

Biochip applications

Devices that use a small, flat substrate that contains biological material are commonly called biochips (literally, biological substrates). Biochips are used for obtaining genetic information. A biochip is a type of miniaturized equipment that integrates tens of thousands of probes made up of genetic material having a known sequence. When the probes are placed in contact with a biological sample (such as from a patient or experiment), only the nucleotide chains complementary to those of the chip hybridize. This action produces a characteristic pattern light, which is read with a scanner and interpreted by a computer.

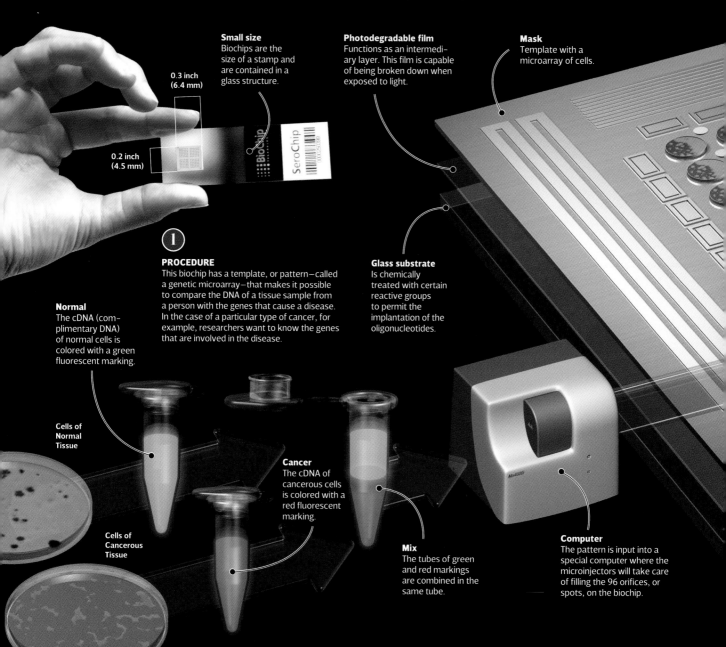

Small size
Biochips are the size of a stamp and are contained in a glass structure.

0.3 inch
(6.4 mm)

0.2 inch
(4.5 mm)

Photodegradable film
Functions as an intermediary layer. This film is capable of being broken down when exposed to light.

Mask
Template with a microarray of cells.

1

PROCEDURE
This biochip has a template, or pattern—called a genetic microarray—that makes it possible to compare the DNA of a tissue sample from a person with the genes that cause a disease. In the case of a particular type of cancer, for example, researchers want to know the genes that are involved in the disease.

Glass substrate
Is chemically treated with certain reactive groups to permit the implantation of the oligonucleotides.

Normal
The cDNA (complimentary DNA) of normal cells is colored with a green fluorescent marking.

Cells of
Normal
Tissue

Cancer
The cDNA of cancerous cells is colored with a red fluorescent marking.

Cells of
Cancerous
Tissue

Mix
The tubes of green and red markings are combined in the same tube.

Computer
The pattern is input into a special computer where the microinjectors will take care of filling the 96 orifices, or spots, on the biochip.

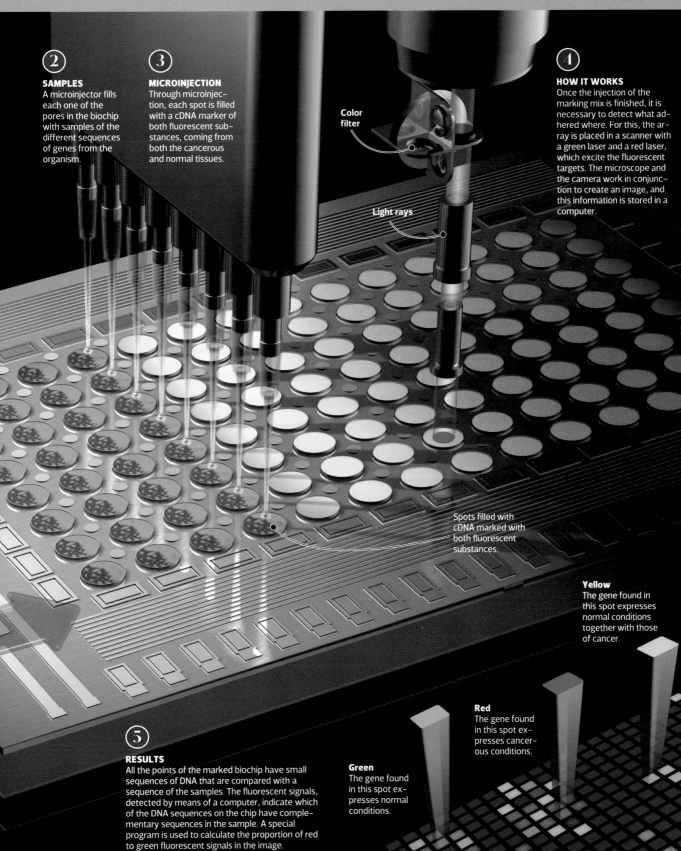

②

SAMPLES
A microinjector fills each one of the pores in the biochip with samples of the different sequences of genes from the organism.

③

MICROINJECTION
Through microinjection, each spot is filled with a cDNA marker of both fluorescent substances, coming from both the cancerous and normal tissues.

Color filter

Light rays

④

HOW IT WORKS
Once the injection of the marking mix is finished, it is necessary to detect what adhered where. For this, the array is placed in a scanner with a green laser and a red laser, which excite the fluorescent targets. The microscope and the camera work in conjunction to create an image, and this information is stored in a computer.

Spots filled with cDNA marked with both fluorescent substances.

Yellow
The gene found in this spot expresses normal conditions together with those of cancer.

Red
The gene found in this spot expresses cancerous conditions.

⑤

RESULTS
All the points of the marked biochip have small sequences of DNA that are compared with a sequence of the samples. The fluorescent signals, detected by means of a computer, indicate which of the DNA sequences on the chip have complementary sequences in the sample. A special program is used to calculate the proportion of red to green fluorescent signals in the image.

Green
The gene found in this spot expresses normal conditions.

Underwater archeology

The invention of autonomous diving equipment enabled underwater archeology to make huge advances in the last century. Underwater archeology has also benefited from collaborations with other specialists, such as geologists, restoration professionals, chemists, and documentary specialists. This specialization helps to provide new information about past battles, ancient submerged ruins the trade of other eras, and even the lives of the crew and passengers of shipwrecked vessels.

The objective

This discipline focuses on the recovery of the long-submerged remains of ships. The main objective is to locate vessels of any kind that have shipwrecked, resulting in numerous archeological remains dispersed across the seabed.

→

Shipwreck
A diver recovers wreckage from a ship that sank in 1025 in the Bay of Serçe Limani (Turkey).

Favorable conditions
Due to the low level of oxygen and reduced action of chemicals in water, non-metallic archeological remains are well-preserved in this medium.

ASSISTIVE TECHNOLOGY

To explore places inaccessible to divers due to depth or lack of space, small remotely operated submarines are used, such as the Triggerfish (at right), equipped with two halogen lamps of 150 watts each with a 984 ft (300 m) range video camera.

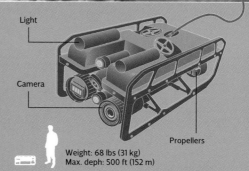

Light

Camera

Propellers

Weight: 68 lbs (31 kg)
Max. deph: 500 ft (152 m)

→ PRESERVATION OF THE FINDINGS The major challenge faced by underwater archaeology is to conserve the remains that are found. When objects are taken from the sea, their balance with the environment is broken, sparking physiochemical processes that can accelerate the decomposition process. It is therefore essential that samples are transported to a restoration laboratory by specialists.

Finding the *Titanic*
The bell from the legendary *Titanic*, which sank in 1912 in the northern Atlantic Ocean. Its wreck was discovered in 1985 by Robert Ballard.

Selection
It is important that only objects that can be preserved in air be removed from the water. Objects that will decompose rapidly in air should be left where they are.

ORGANIZATION AND TOOLS

Tasks are usually distributed between four teams: one on the surface, in a vessel vertically aligned with the site; the auxiliary team, at a nearby beach, controlling logistics; the receiving team, for preserving objects; and the sub-aquatic team, made up of divers who explore the site. These are some tools used by divers:

Notebook
On site records are essential. Special pens and laminated paper on which the pens can write are used for taking notes.

Lifting bag
Remains (or sediments) that are too large to be carried by the diver are raised using a lifting device.

Grid
A portable grid enables the archeological site to be sectioned and systematically cleaned and surveyed with either sketches or photographs.

Airlift tube
A long tube, controlled from the surface, which sucks up water, removing the sediments around the archeological remains and cleaning the area.

Transplants

When all other options for treating a disease run out, the last place to turn is often to replace a sick organ with a healthy one through transplantation. Some organs can be donated from a living person, as with kidneys, where the procedure does not harm the donor. Often however, organs are harvested from a donor corpse. Advances are constantly being made and the most recent transplantation is the face transplant, which involves working with many highly complex nerves.

Face transplant

The surgery performed to replace a damaged face, usually caused by burns or degenerative diseases, is a cutting-edge, complicated technique. This surgery has a prolonged recovery time.

➜

The nerves
The nerves can only be joined through microsurgery. The operation is very complicated because the face is full of nerve endings.

① REMOVAL
The patient's face is removed. Transplant can be partial or total.

② PREPARATION
As the face is a complex network of blood vessels, capillaries, arteries and veins, it is carefully prepared.

③ ALIGNMENT
Doctors align the new face to the patient and vessels and nerves are connected to the new tissue.

④ RECOVERY
The skin is sutured and a period of adaptation of the skin to the new body begins. The patient has to integrate the new face into his or her body image.

TYPES OF TRANSPLANTS

Of the two types of transplant operations (organs and tissues), organ transplants are by far more difficult. They require complex surgeries to achieve the splicing of vessels and ducts. Tissue transplants are simpler: cells are injected, to be implanted later.

Allograft
Consists of the donation of organs from one individual to another genetically different individual of the same species.

Autograft
A transplant in which the donor and the recipient are the same person. The typical case is a skin graft from a healthy site to an injured one.

Isograft
A transplant in which the donor and the recipient are genetically identical.

Xenograft
A transplant in which the donor and the recipient are of different species (e.g., from a monkey to a human). This type generates the strongest potential rejection response by the body of the recipient.

↓ HEART TRANSPLANT

Once the patient is anesthetized, an incision is made in his chest. Heart and lung functions are replaced by a mechanical pump and the aorta is clamped. That is when doctors change the heart. After the transplant, the healthy heart is stitched to the patient's arteries and veins, the aortic clamp is released and all bleeding is kept under control.

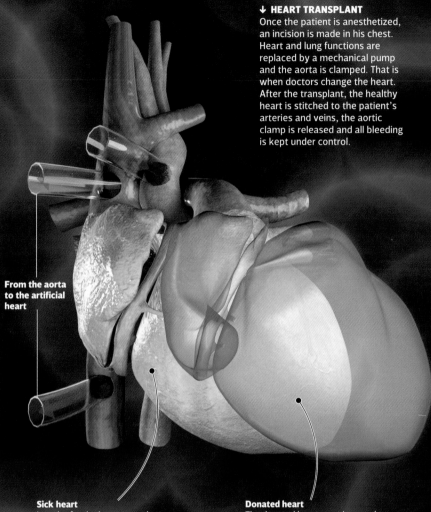

From the aorta to the artificial heart

Donors

While most donations occur after the donor has died, some others such as blood, umbilical cord blood, or kidney can be donated while the donor is alive.

Sick heart
In order for the heart transplant to take place, the heart must be stopped. This is achieved by reducing the body temperature of the patient. This has the benefit of also preserving the circulation of blood through the brain.

Donated heart
The donated heart must be an adequate size, taking into account the beneficiary's needs. In general, when a donor is of average height and weight, his or her heart most probably will work well on the majority of heart-transplant beneficiaries.

LIVER TRANSPLANT

People who suffer advanced, ir-reversible, life-threatening hepatic conditions now have the possibility of attempting a liver transplant. The most typical liver transplant cases are those of people who suffer chronic hepatitis or primary biliary cirrhosis, an autoimmune disease. Donors must not be infected in any way and cannot be suffering from any cardiac or pulmonary disease at the time of donation.

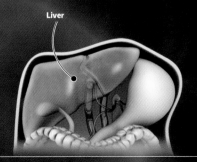

Liver

① Donated liver
The organ, along with all its blood vessels and its bile duct, is removed immediately after the death of the donor.

② The new liver
is fused with the vena cava and the rest of the blood vessels. The opposite ends of the bile duct are sutured. A probe is inserted inside the reconstructed bile duct to drain the blood and the bile.

4D Ultrasound

The 4D ultrasound is the latest improvement in obstetric diagnostic examinations. Ultrasound imaging in four dimensions incorporates time as a new variable, and it produces color images in real time that give the impression of watching a movie of a baby as it grows inside the uterus. However, it is not a movie, properly speaking but the sweep of ultrasonic waves that are reflected as echoes by the fetus. These echoes are analyzed and converted into images by powerful processors that perform mathematical calculations. The use of 4D ultrasound has not yet been completely embraced by doctors, many of whom continue to prefer traditional two-dimensional ultrasounds.

20 to 20,000 hertz

The times per second that the transducer emits ultrasonic waves and detects the waves that are reflected by the fetus.

5,000

The times per second that the transducer emits ultrasonic waves and detects the waves that are reflected by the fetus.

The ultrasonic window

The ultrasound machine uses a handheld probe that is moved over the mother's abdomen. The probe contains transducers that emit ultrasonic (high-frequency) waves that pass through the abdomen and bounce off the baby, creating echoes. These reflected waves are detected by the transducer and then converted into images.

→

HOW IT WORKS

Although the result of the exam is a moving image of a fetus in color, the ultrasound machine does not use optical equipment but only sound waves reflected by the baby. This imaging method is generally not thought to pose a risk for either the fetus or the mother.

2

ECHO

The ultrasonic waves collide with and bounce off of fetal tissue. The frequencies used are inaudible to the human ear.

1

EMISSION

The transducer emits ultrasonic waves at specific frequencies that will pass through external tissue into the uterus, where the baby is. A motor varies the plane of the emitted waves many times a second to produce three-dimensional images.

3

RECEPTION

The transducer receives the waves reflected from the tissues of the fetus. Depending on their characteristics and how they were modified, the processor extracts information from the reflected waves and converts them into moving images in real time.

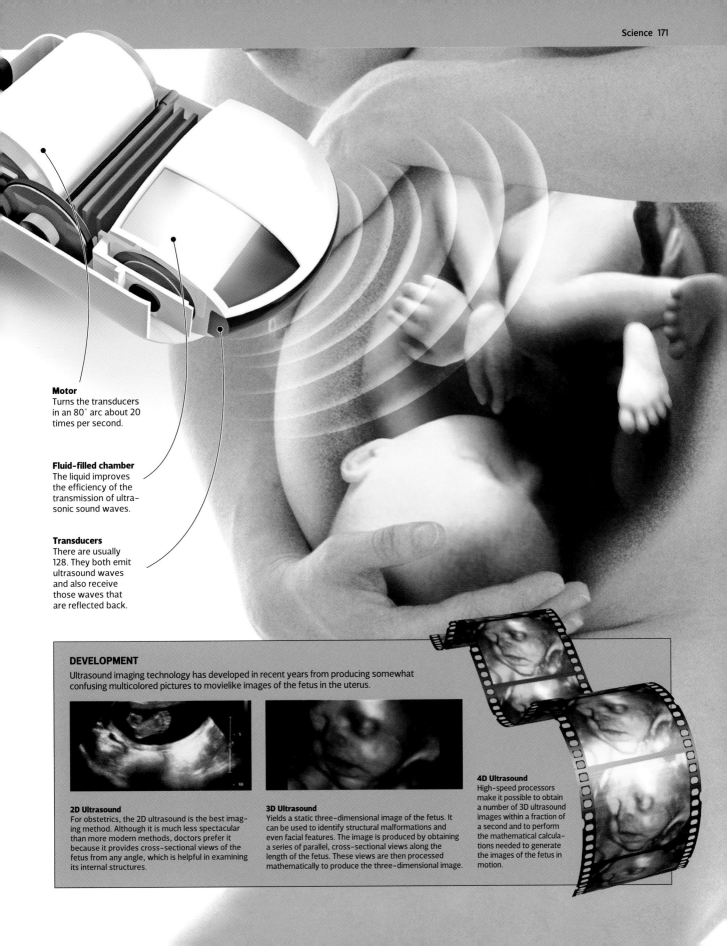

Motor
Turns the transducers in an 80° arc about 20 times per second.

Fluid-filled chamber
The liquid improves the efficiency of the transmission of ultra-sonic sound waves.

Transducers
There are usually 128. They both emit ultrasound waves and also receive those waves that are reflected back.

DEVELOPMENT

Ultrasound imaging technology has developed in recent years from producing somewhat confusing multicolored pictures to movielike images of the fetus in the uterus.

2D Ultrasound
For obstetrics, the 2D ultrasound is the best imaging method. Although it is much less spectacular than more modern methods, doctors prefer it because it provides cross-sectional views of the fetus from any angle, which is helpful in examining its internal structures.

3D Ultrasound
Yields a static three-dimensional image of the fetus. It can be used to identify structural malformations and even facial features. The image is produced by obtaining a series of parallel, cross-sectional views along the length of the fetus. These views are then processed mathematically to produce the three-dimensional image.

4D Ultrasound
High-speed processors make it possible to obtain a number of 3D ultrasound images within a fraction of a second and to perform the mathematical calcula-tions needed to generate the images of the fetus in motion.

In vitro fertilization

Ever since the first successful case of in vitro fertilization in the United Kingdom almost three decades ago, this technique has become the most popular and widespread method of assisted reproductive technology. It involves removing a woman's ova, or eggs, and fertilizing them outside the woman's womb; in fact, the procedure is done in a laboratory, to avoid various problems that can hinder a natural pregnancy. Once fertilized, the embryo is implanted in the uterus to continue gestation.

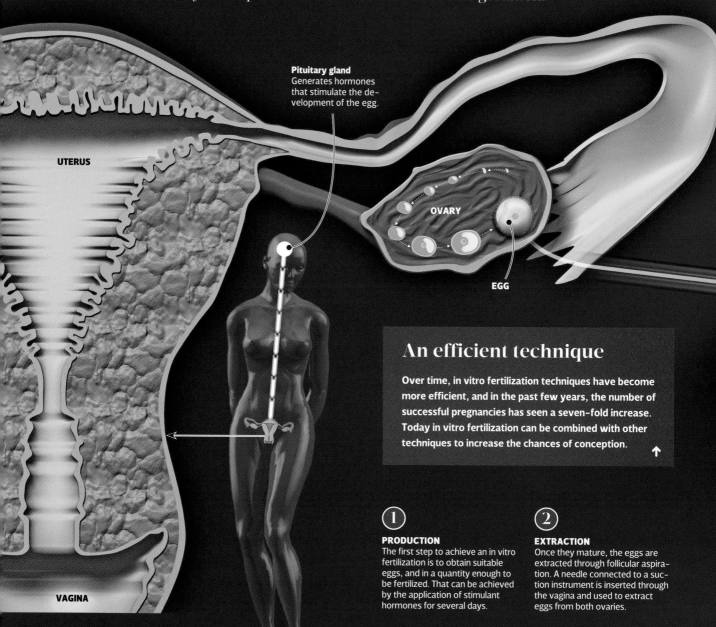

Pituitary gland
Generates hormones that stimulate the development of the egg.

UTERUS

OVARY

EGG

VAGINA

An efficient technique

Over time, in vitro fertilization techniques have become more efficient, and in the past few years, the number of successful pregnancies has seen a seven-fold increase. Today in vitro fertilization can be combined with other techniques to increase the chances of conception. ↑

① PRODUCTION
The first step to achieve an in vitro fertilization is to obtain suitable eggs, and in a quantity enough to be fertilized. That can be achieved by the application of stimulant hormones for several days.

② EXTRACTION
Once they mature, the eggs are extracted through follicular aspiration. A needle connected to a suction instrument is inserted through the vagina and used to extract eggs from both ovaries.

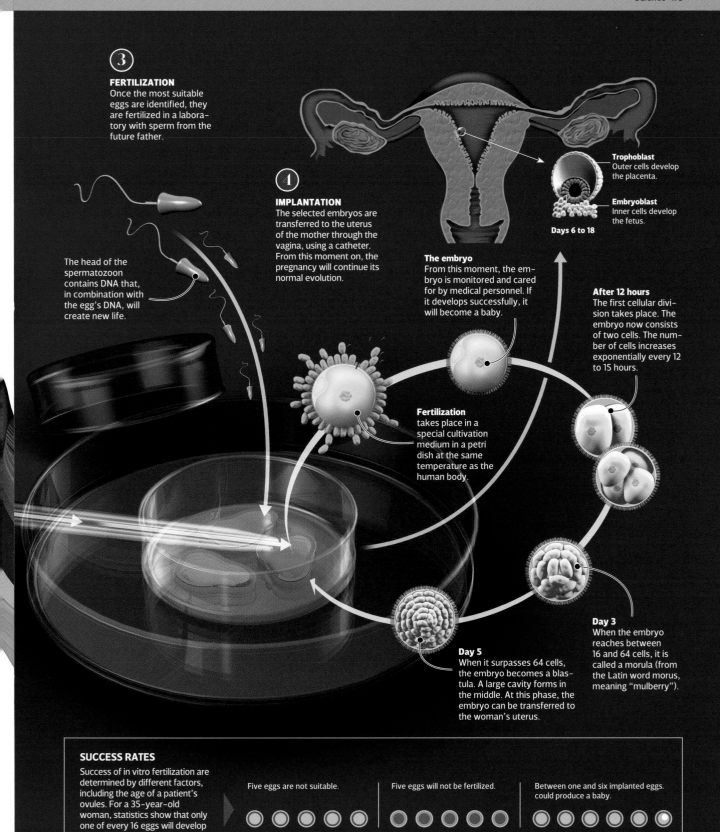

③ **FERTILIZATION**
Once the most suitable eggs are identified, they are fertilized in a laboratory with sperm from the future father.

The head of the spermatozoon contains DNA that, in combination with the egg's DNA, will create new life.

④ **IMPLANTATION**
The selected embryos are transferred to the uterus of the mother through the vagina, using a catheter. From this moment on, the pregnancy will continue its normal evolution.

The embryo
From this moment, the embryo is monitored and cared for by medical personnel. If it develops successfully, it will become a baby.

Trophoblast
Outer cells develop the placenta.

Embryoblast
Inner cells develop the fetus.

Days 6 to 18

After 12 hours
The first cellular division takes place. The embryo now consists of two cells. The number of cells increases exponentially every 12 to 15 hours.

Fertilization
takes place in a special cultivation medium in a petri dish at the same temperature as the human body.

Day 3
When the embryo reaches between 16 and 64 cells, it is called a morula (from the Latin word morus, meaning "mulberry").

Day 5
When it surpasses 64 cells, the embryo becomes a blastula. A large cavity forms in the middle. At this phase, the embryo can be transferred to the woman's uterus.

SUCCESS RATES
Success of in vitro fertilization are determined by different factors, including the age of a patient's ovules. For a 35-year-old woman, statistics show that only one of every 16 eggs will develop and result in a pregnancy.

Five eggs are not suitable.

Five eggs will not be fertilized.

Between one and six implanted eggs. could produce a baby.

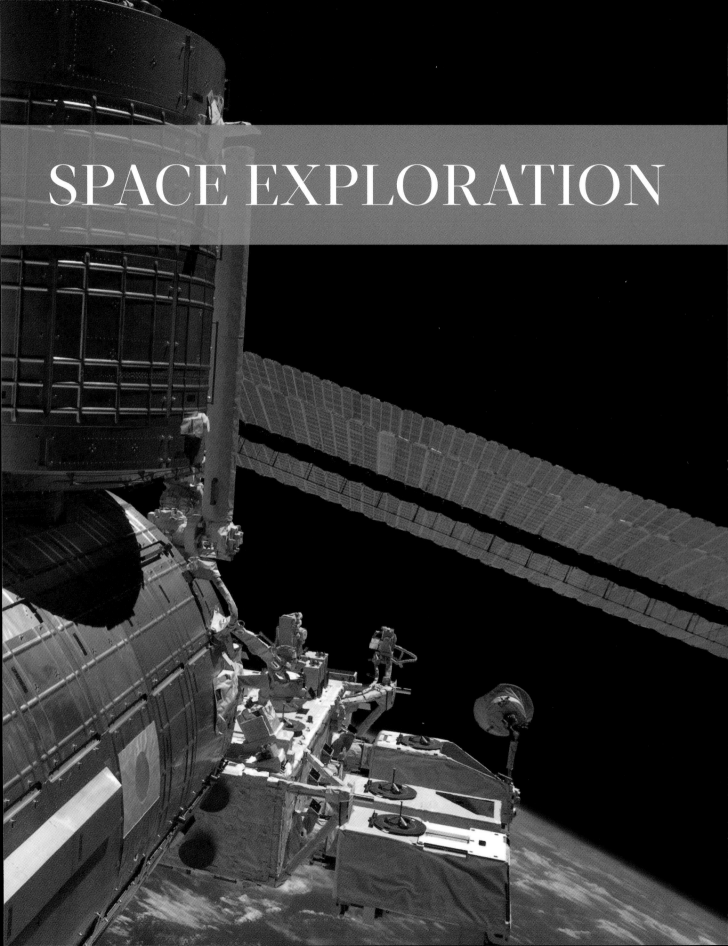

SPACE EXPLORATION

Rockets

Developed in the first half of the 20th century, rockets are necessary for sending any kind of object into space. They produce enough force to leave the ground with cargo and, in a short period of time, they acquire the speed necessary to escape gravity completely and orbit around the Earth. Approximately, more than one rocket per week is launched into space from somewhere in the world.

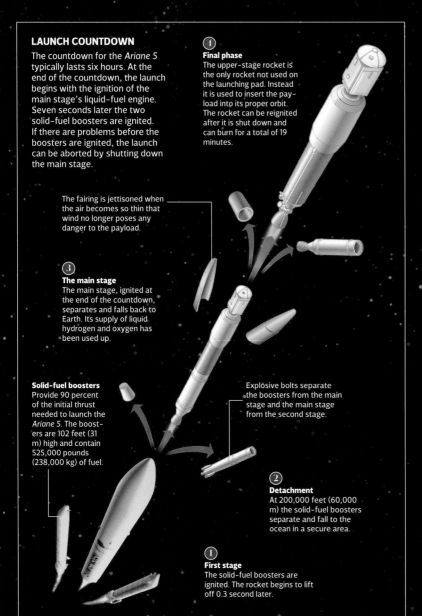

LAUNCH COUNTDOWN
The countdown for the *Ariane 5* typically lasts six hours. At the end of the countdown, the launch begins with the ignition of the main stage's liquid-fuel engine. Seven seconds later the two solid-fuel boosters are ignited. If there are problems before the boosters are ignited, the launch can be aborted by shutting down the main stage.

④ Final phase
The upper-stage rocket is the only rocket not used on the launching pad. Instead it is used to insert the payload into its proper orbit. The rocket can be reignited after it is shut down and can burn for a total of 19 minutes.

The fairing is jettisoned when the air becomes so thin that wind no longer poses any danger to the payload.

③ The main stage
The main stage, ignited at the end of the countdown, separates and falls back to Earth. Its supply of liquid hydrogen and oxygen has been used up.

Solid-fuel boosters
Provide 90 percent of the initial thrust needed to launch the *Ariane 5*. The boosters are 102 feet (31 m) high and contain 525,000 pounds (238,000 kg) of fuel.

Explosive bolts separate the boosters from the main stage and the main stage from the second stage.

② Detachment
At 200,000 feet (60,000 m) the solid-fuel boosters separate and fall to the ocean in a secure area.

① First stage
The solid-fuel boosters are ignited. The rocket begins to lift off 0.3 second later.

Spaceflights

Access to space—whether for placing satellites into orbit, sending probes to other planets, or launching astronauts into space—has become almost routine and is big business for countries that have launch capabilities.

→

363 feet
(111 m)

The height of the *Saturn V*, the largest rocket ever launched. It was used in the late 1960s and early 1970s to take astronauts to the Moon. During a launch, it could be heard 90 miles (150 km) away.

Thermal insulation
To protect the combustion chamber from the high temperatures of the burning fuel, the walls are sprayed with rocket fuel. This process manages to cool the engine off.

Thrusters
Expel gases so that the rocket can begin its ascent.

Conical nose cone
Protects the cargo.

Upper payload
Up to two satellites.

Lower payload
Carries up to two satellites.

Upper engines
Release the satellites at a
precise angle and speed.

Liquid oxygen tank
Contains 286,000
pounds (130,000 kg) for
combustion.

Liquid hydrogen tank
Contains 225 tons.

Booster rockets
Burn fuel for two
minutes.

Main engine
Burns for 10 minutes.

↓ LAUNCH WINDOW

Rockets must be launched at predetermined times,
which depend on the objective of the launch. If the
objective is to place a satellite into orbit, the latitude
of the launched rocket needs to coincide with the
trajectory of the desired orbit. When the mission
involves docking with another object in space, the
launch window might fall within only a few minutes.

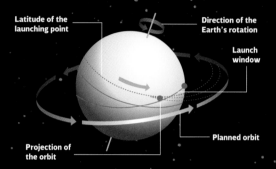

Latitude of the launching point

Direction of the Earth's rotation

Launch window

Projection of the orbit

Planned orbit

HOW IT WORKS

To do its job, the rocket must
overcome gravity. As it rises, the
mass of the rocket is reduced
through the burning of its fuel.
Moreover, as the distance from
the Earth increases, the effect of
gravity decreases.

Action and reaction
The thrust of the
rocket is the reaction
resulting from the
action of the hot
exhaust escaping
from the rocket.

Rocket's thrust

Earth's gravity

HOW IT FLIES

The hot gases produced
by the burning fuel push in
all directions. As the gases
escape through the open
nozzle, they generate an
opposing force.

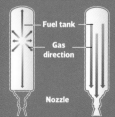

Fuel tank

Gas direction

Nozzle

FLIGHT GUIDANCE

The rocket's guidance
computer uses data
from laser gyroscopes to
control the inclination of
the nozzles, directing the
rocket along its proper
flight path.

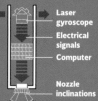

Laser gyroscope

Electrical signals

Computer

Nozzle inclinations

Space Observatory

Thanks to the data sent in 2001 from the NASA observatory WMAP (Wilkinson Microwave Anisotropy Probe), scientists have managed to make the first detailed maps of cosmic background radiation. Cosmic background radiation is thought to be the echoes of the Big Bang. Experts believe that this map reveals clues about when the first generation of stars was formed.

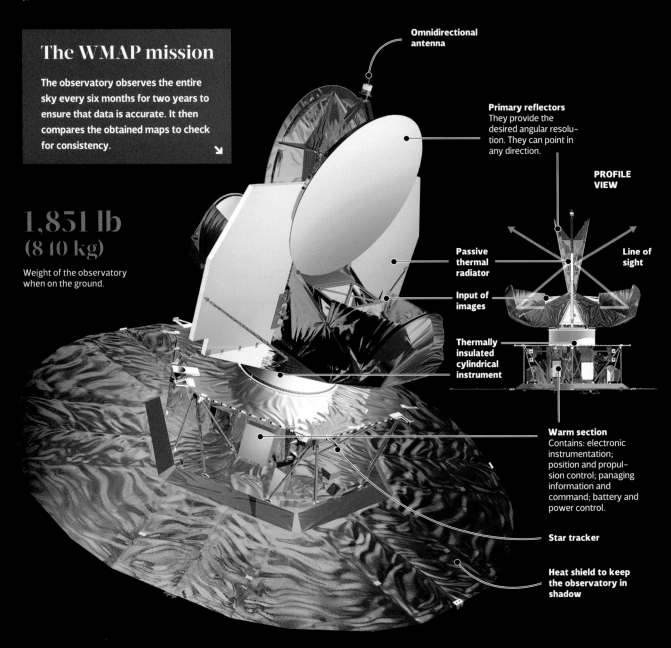

The WMAP mission

The observatory observes the entire sky every six months for two years to ensure that data is accurate. It then compares the obtained maps to check for consistency. ↘

1,851 lb
(810 kg)

Weight of the observatory when on the ground.

Omnidirectional antenna

Primary reflectors
They provide the desired angular resolution. They can point in any direction.

PROFILE VIEW

Line of sight

Passive thermal radiator

Input of images

Thermally insulated cylindrical instrument

Warm section
Contains: electronic instrumentation; position and propulsion control; panaging information and command; battery and power control.

Star tracker

Heat shield to keep the observatory in shadow

OBSERVATION

In order to observe the whole sky, the probe is located at the so-called L2 Lagrange Point, 0.9 million miles (1.5 million km) from Earth. This point provides a stable environment, away from the influence of the sun. WMAP observes the sky at different stages and measures temperature differences between different cosmic regions. Every six months it completes a full sky coverage.

② DAY 90 (3 MONTHS)

The probe has completed coverage of half the sky. Each hour it covers a sector of 22.5°.

WMAP TRAJECTORY

Before heading to the L2 Lagrange Point, the probe performed a flyby of the moon, using lunar gravity to propel it towards L2.

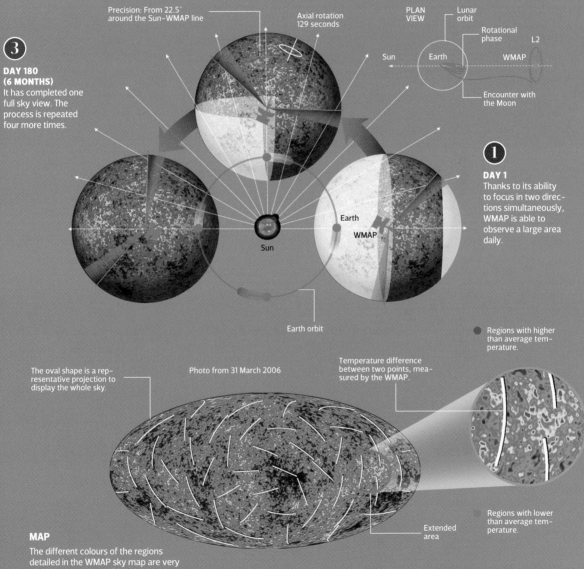

Precision: From 22.5° around the Sun-WMAP line

Axial rotation 129 seconds

PLAN VIEW

Lunar orbit

Rotational phase

L2

Sun

Earth

WMAP

Encounter with the Moon

③ DAY 180 (6 MONTHS)

It has completed one full sky view. The process is repeated four more times.

Earth

WMAP

Sun

① DAY 1

Thanks to its ability to focus in two directions simultaneously, WMAP is able to observe a large area daily.

Earth orbit

Regions with higher than average temperature.

The oval shape is a representative projection to display the whole sky.

Photo from 31 March 2006

Temperature difference between two points, measured by the WMAP.

Regions with lower than average temperature.

Extended area

MAP

The different colours of the regions detailed in the WMAP sky map are very slight temperature differences in the cosmic microwave background. This radiation, remnants of the Big Bang, was discovered 40 years ago, but can only now be described in detail.

COBE, THE PREDECESSOR

COBE's results from 1989 provided the kick-start for the WAMP project. The resolution was much lower, so the spots are larger.

Space Shuttle

Unlike conventional rockets, the space shuttles were used over and over again to put satellites into orbit. Until 2011, these vehicles were used to launch and repair satellites and as astronomical laboratories. The American fleet had five space shuttles over its history: *Challenger* and *Columbia* (exploded in 1986 and 2003, respectively), *Discovery*, *Atlantis*, and *Endeavour*. The space shuttle program ended in 2011 with the retirement of the three remaining shuttles.

Reusable

The space shuttle was the first space-craft capable of returning to Earth on its own being used in multiple missions. The orbiter played a key role in building the International Space Station. →

Satellite
Stays in the cargo hold and is moved by the arm.

Mechanical arm
Moves satellites in and out of the cargo module.

Command cabin

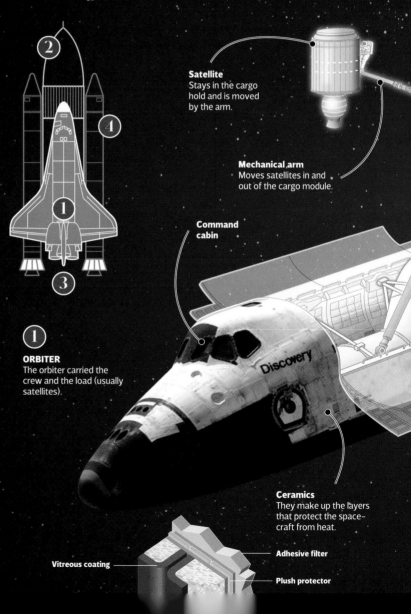

ORBITER
The orbiter carried the crew and the load (usually satellites).

Discovery

Ceramics
They make up the layers that protect the space-craft from heat.

Vitreous coating

Adhesive filter

Plush protector

↓ **CABIN** Divided into two levels: an upper one for the pilot and co-pilot (and up to two astronauts), and a lower one where everyday work is done. The habitable volume of the cabin is 2,472 ft³ (70 m³).

Controls
In the cockpit there are more than 2,000 separate controls.

Control keypad

Pilot's seat

Co-pilot's seat

②

EXTERNAL FUEL TANK
Connects the shuttle to the launcher rockets. Carries loads of liquid oxygen and hydrogen, which are combusted through a tube connecting each container with the next. The tank is lost on each trip.

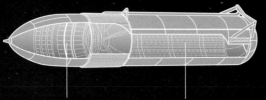

Liquid oxygen

Liquid hydrogen

③

THE MAIN ENGINES
There are three of these, which feed liquid oxygen and hydrogen from the external tank. Each engine has a controller based on a digital computer, which makes adjustments the thrust and correct the fuel mixture.

Circulation of
liquid hydrogen

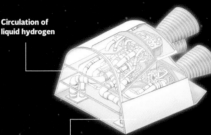

Heat shield

Vertical wing
The vertical wing was used during the descent to stabilize and to control the direction.

Orbital motors
Provide the thrust for entering orbit and orbital adjustments that may be needed. They are located on the outside of the fuselage.

Ignition section

④

SOLID ROCKETS
They are designed to last about 20 flights. After each trip they are retrieved from the ocean and refurbished. They take the shuttle to an altitude of 27 miles (44 km) and can support the full weight of the shuttle.

Delta wings
The shuttle had to descend, gliding without fuel. So its wings resembled those of a paper airplane.

Solid fuel

Thruster mouth

United States

Gates
They open when the device reaches low Earth orbit. They are thermal panels that protect the spacecraft from overheating.

LANDING SEQUENCE

①
Slowdown
The shuttle turned and activated reverse thrusters to slow down.

②
Entering the atmosphere
The high speed generated very high temperatures. The shuttle had to consume all of its fuel before reentering the atmosphere to avoid explosion.

③
Downward spiral
The remainder of the descent was made by gliding on its delta wings, without any engine.

⑤
Slows the descent by parachute

①
Approaches the runaway

NASA

The National Astronautics and Space Administration (NASA) is the US space agency. It was created in 1958 as part of the "space race" being contested with the Soviet Union, and it coordinated all national activities related to space exploration. NASA has facilities throughout the country, including a the Kennedy Space Center, the main launching center.

Control from Earth

Monitoring of astronauts' activity is done from operations centers. In the United States, NASA coordinates manned missions from the Mission Control Center located in the Johnson Space Center in Houston. The unmanned missions are supervised from the Jet Propulsion Laboratory in Los Angeles.

Screen 1
Records the location of satellites and other objects in orbit.

CONSOLE

The Operations Control Room contains about a hundred consoles. The consoles form desks with an area for more than one monitor. They have drawers and counters for providing a working area.

Folding table
For supporting objects and books.

Monitor
To display data from spacecraft and other systems.

Flight direction (Row 3)
Responsible for the countdown before liftoff and the flight plans.

FLIGHT DIRECTOR

Directorate (Row 4)
The lead authorities are located in the fourth row, from which they coordinate the crew's flight operations.

Rear sliding drawer
To keep information and papers.

Protective covering
Prevents damage to the console system.

THE BIG SCREEN

An enormous screen dominates the Operations Control Center. It provides information on the location and orbital trajectory of spacecraft in flight, as well as other data. The screen is of vital importance for the operators, because it allows for the rapid reading of information to take action efficiently and prevent accidents.

Screen 2
Shows the location and path of spacecraft in orbit.

Liftoff monitor (Row 1)
Controls the trajectory and carries out course adjustments to the spacecraft.

Medical section (Row 2)
The second row checks the astronauts' health and maintains contact with the crew.

NASA BASES

NASA has facilities throughout the United States that develop and research, flight simulation and astronaut training. NASA's headquarters are in Washington, D.C., and the Flight Control Center is in Houston.

Ames Research Center

Lyndon B. Johnson Control Center

Marshall Space Flight Center

Glenn Research Center

Goddard Institute for Space Studies

Independent Verification and Validation Facility

Langley Research Center

NASA CONTROL CENTER
Washington, D.C.

Wallops Flight Facility

Jet Propulsion Laboratory
Designs flight systems and provides technical advice.

White Sands Test Facility

Dryden Flight Research Center
Responsible for operations related to the atmosphere. Has been in operation since 1947.

Michoud Assembly Facility

John F. Kennedy Space Center

Goddard Space Flight Center
Designs, manufactures and monitors scientific satellites to investigate the Earth and other planets.

OPS PLANNER

X-ray observatories

In July 1999, the Chandra Observatory was put into orbit. This telescope can view the heavens using X-rays with an angular resolution of 0.5 arc-seconds, making it one thousand times more powerful than the first orbital X-ray telescope, named Einstein. This feature allows it to detect light sources that are 20 times more diffuse. The group tasked with constructing the X-ray telescope was responsible for developing technologies and processes that had never been applied.

Cutting-edge technology

The satellite system provides the structure and equipment required for the telescope and scientific instruments to work as an observatory. To control the critical temperatures of its components, Chandra has a special system comprising radiators and thermostats. The satellite's electricity is supplied by solar panels and is stored in three batteries.

HOW THE IMAGE IS CREATED

The information gathered by Chandra is extracted into images and tables with coordinates on the X- and Y-axes.

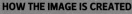

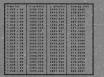

① Table
Contains the time, position and energy collected bt Chandra's observations.

② X-axis
The data extends horizontally through the grid.

③ Y-axis
The data extends vertically through the grid.

① OBSERVATION
The telescope's camera takes an X-ray image and sends it to the Deep Space Network for processing.

Solar panel

Photographic camera

High-resolution mirror

4 hierarchical hyperboloids

X-rays

④ CHANDRA X-RAY CONTROL CENTRE
Tasked with ensuring the observatory's functionality and with receiving images. The operators are also responsible for preparing commands, determining the altitude and monitoring the condition and safety of the satellite.

③ JET PROPULSION LABORATORY
Information is received from the Deep Space Network and processed.

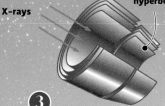

5 years

The already-surpassed life expectancy of the Chandra observatory mission.

33 ft
(10 m)

Solar panel

Transmission grids

Optical array

Scientific instrument module

High-resolution camera

Low gain antenna

Every 8 hours

Chandra contacts the Deep Space Network.

DEEP SPACE NETWORK

This network is used to communicate with the spacecraft and to receive information.

DEEP SPACE NETWORK

NASA's international network of antennae, which support interplanetary missions orbiting Earth and radio-astronomy missions, has three complexes, each of which have at least four Deep Space stations, equipped with ultra-sensitive receiver systems and large-scale parabolic antennae.

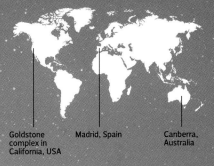

Goldstone complex in California, USA

Madrid, Spain

Canberra, Australia

THE ANTENNAE

Each complex has a system comprising at least four antennae.

❶ Antenna measuring 85 ft (26 m) in diameter.

❷ Low gain antenna measuring 112 ft (34 m) in diameter.

❸ Antenna measuring 230 ft (70 m) in diameter.

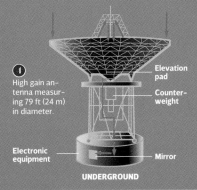

❹ High gain antenna measuring 79 ft (24 m) in diameter.

Elevation pad

Counterweight

Electronic equipment

Mirror

UNDERGROUND

Space Probes

Since the first spacecraft in the 1960s, the contribution to science made by space probes has been considerable. Mostly solar-powered, these unmanned machines are equipped with sophisticated instruments that make it possible to study planets, moons, comets, and asteroids in detail. One particularly renowned probe is the Mars Reconnaissance Orbiter (MRO), which was launched to study Mars in 2005.

APPROACH TO MARS

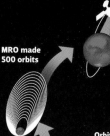

MRO made 500 orbits

Ⓒ **Final orbit**
It travelled along an almost circular orbit, suitable for obtaining data.

Orbit

Ⓑ **Braking**
To get closer to the planet, the space-craft slowed down over a six-month period.

Mars

Ⓐ **Start**
The probe's first orbit travelled along an enormous ellipti-cal path.

Mars Reconnaissance Orbiter

The main objective of this orbiting probe is to seek out traces of water on the surface of Mars. The probe was launched in summer 2005 by NASA and reached Mars on 10 March 2006. It travelled 116 million km (72 million miles) in seven months. NASA plans to keep using it past the mid-2020s.

Mar's orbit

Earth's orbit

Sun

Mars

Earth

① **Launch**
Took place on Au-gust 12, 2005, from Cape Canaveral, USA.

② **Cruising**
The probe travelled for seven and a half months before reaching Mars.

③ **Path correction**
Four maneuvers were made to en-sure the correct orbit was reached.

④ **Arrival on Mars**
In March 2006, MRO passed into the southern hemisphere of Mars. The probe slowed down considerably.

⑤ **Scientific phase**
The probe began its analysis phase on the surface of Mars. It found evidence of water.

72 million miles
(116 million km)

Travelled by the Mars Reconnaissance Orbiter on its journey to Mars.

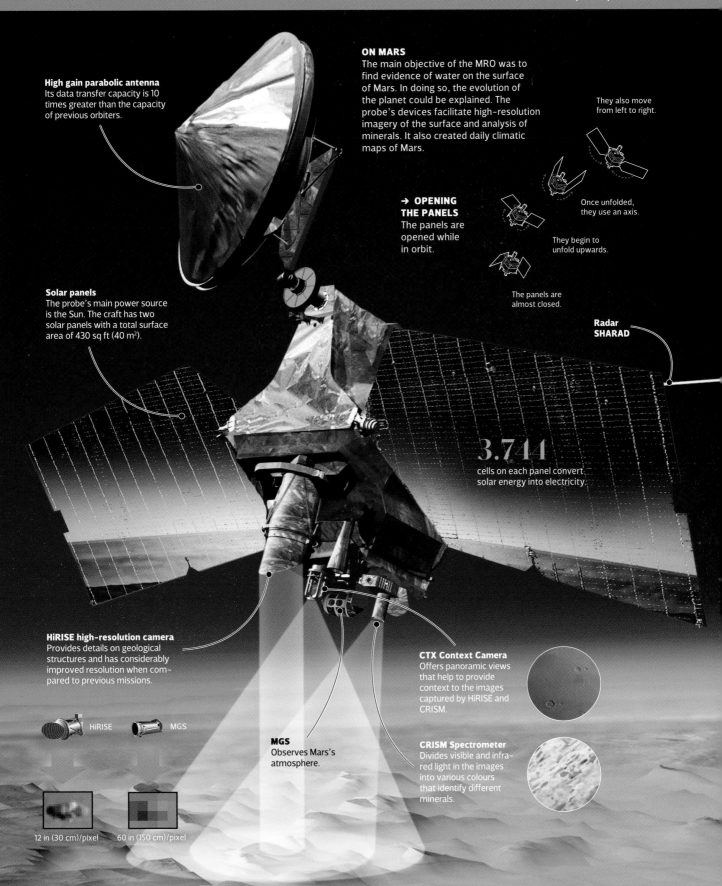

High gain parabolic antenna
Its data transfer capacity is 10 times greater than the capacity of previous orbiters.

ON MARS
The main objective of the MRO was to find evidence of water on the surface of Mars. In doing so, the evolution of the planet could be explained. The probe's devices facilitate high-resolution imagery of the surface and analysis of minerals. It also created daily climatic maps of Mars.

They also move from left to right.

→ **OPENING THE PANELS**
The panels are opened while in orbit.

Once unfolded, they use an axis.

They begin to unfold upwards.

The panels are almost closed.

Solar panels
The probe's main power source is the Sun. The craft has two solar panels with a total surface area of 430 sq ft (40 m²).

Radar SHARAD

3.744
cells on each panel convert solar energy into electricity.

HiRISE high-resolution camera
Provides details on geological structures and has considerably improved resolution when compared to previous missions.

HiRISE   MGS

CTX Context Camera
Offers panoramic views that help to provide context to the images captured by HiRISE and CRISM.

MGS
Observes Mars's atmosphere.

CRISM Spectrometer
Divides visible and infra-red light in the images into various colours that identify different minerals.

12 in (30 cm)/pixel   60 in (150 cm)/pixel

Mars Rovers

Spirit and *Opportunity*, the twin robots launched in June 2003 from Earth, arrived on Mars in January 2004. They were the first robots to travel on the surface of the red planet. Both form part of NASA's Mars Exploration Rovers mission and are equipped with tools to drill rocks and collect samples from the ground for analysis.

Water on Mars

The main objective of the mission was to find evidence of past water activity on Mars. Although the robots have found evidence of this, they have been unable to find living microorganisms, given that the ultraviolet radiation and oxidizing nature of the soil make life on Mars impossible. The question that remains unanswered is whether life may have existed on Mars at some stage in the past. And what's more is whether life currently exists in the subsoil on Mars, where conditions may be more favorable. →

70,000
images obtained by *Spirit* during its first two years.

Photograph of the surface taken by *Spirit*.

80,000
images obtained by *Opportunity* during its first two years.

Footprint and photograph taken by *Opportunity*.

Aeroshell

① Deceleration
81 miles (130 km) from the surface, the aeroshell slows down from 9,941 to 994 mph (16,000 to 1,600 km/h).

Parachute

② Parachute
6.2 miles (10 km) from the surface, the parachute opens to slow down the descent.

③ Descent
The shield that offered the rover heat protection separates from the input module.

Input module

④ Rockets
33–49 ft (10–15 m) from the surface, two rockets are ignited to slow down the descent. Two airbags are then inflated to surround and protect the landing gear.

⑤ Airbags
The landing gear and airbags detach from the parachute and fall to Mars's surface.

Descent rockets

⑥ Landing
The airbags deflate. The "petals" that protect the spacecraft open. The vehicle emerges.

⑦ Instruments
The robot opens its solar panels, the mast camera and its antennae.

Vectran airbags

→ **HOW TO REACH MARS** The journey to Mars took seven months. Once inside Mars's atmosphere, a parachute is deployed to slow down the descent.

**Panoramic
(PANCAM)**

CAMERAS

Two navigational
cameras and two
panoramic cameras
installed on the mast.

45°   PANCAM

16°
0°   Vertical
−16°   viewing
                angles
         NAVCAM
−45°

Navigation
(NAVCAM)

Panoramic
(PANCAM)

**Omni-directional
shortwave antenna**
Transmits the information
gathered by the robot to
the control center on Earth.

Solar panels
Capture solar light
and transform it into
energy. Generates
around 140 watts
every five hours.

Inertial measurement unit
Provides information on its position using the X, Y, and Z axes.

Antenna

**Electronic
module**

**Dual camera
mounted on
the front**

Battery

**Folded
arm**

Bent arm

**X-wave
radio**

Abrasion
tools

Microscope

Mössbauer
spectroscope

X-ray
spectrometer

2 in/sec
(5 cm/sec)

Maximum speed on flat surfaces.

Three petals and a
central base form the
craft's protective shield.

MOVEMENT AND PROPULSION

The robot is equipped with six wheels.
Each one has an individual electric
motor, offering the vehicle excellent
traction capacity.

Operation cycles
The robot is programmed to
work in cycles of 30 seconds.

The propulsion
system enables the
robot to overcome
small obstacles.

Stabilization

Space Stations

Living on space stations makes it possible to study the effects of remaining in outer space for extended periods of time, while providing an environment for scientists to conduct experiments in laboratories. These stations are equipped with systems that provide the crew with oxygen and that filter exhaled carbon dioxide.

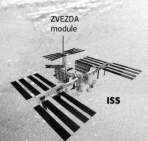

The ISS

The International Space Station (ISS) is the result of the merger of NASA's Freedom project with Mir-2, run by the Russian Federal Space Agency (RKA). Construction began in 1998 and it continues to expand, using modules provided by countries across the globe. Its inhabitable surface area is equal to that of a Boeing 747.

→

↓ PROVISION AND WASTE
The Russian spacecraft ATV connects to the ISS, to provide supplies and remove waste.

ATV   ISS

ZVEZDA MODULE
The main Russian contribution to the station was the first living space. It houses between three and seven astronauts.

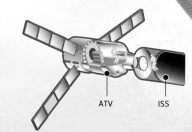

Wardrobe

Beds

Shower

Storage and kitchen

Control and communications area

Connecting node between modules

ZVEZDA module

The floor and roof are different colors, to facilitate orientation.

Unfolding solar panels

ISS

PHASES OF CONSTRUCTION

Zarya module (November 1998)
First sector put into orbit. It powered the first construction stages of the ISS.

Unity module (December 1998)
Connection finger between the living and working area modules. Provided by the EU.

Zvezda module (July 2000)
The structural and functional heart of the ISS. Fully built and put into orbit by Russia.

Z1 Truss and Ku-Band antenna (October 2000)
Neutralizes the static electricity generated in the ISS and facilitates communication with Earth.

P6 Truss (November 2000)
Structural module that features radiators to disperse the heat generated in the station.

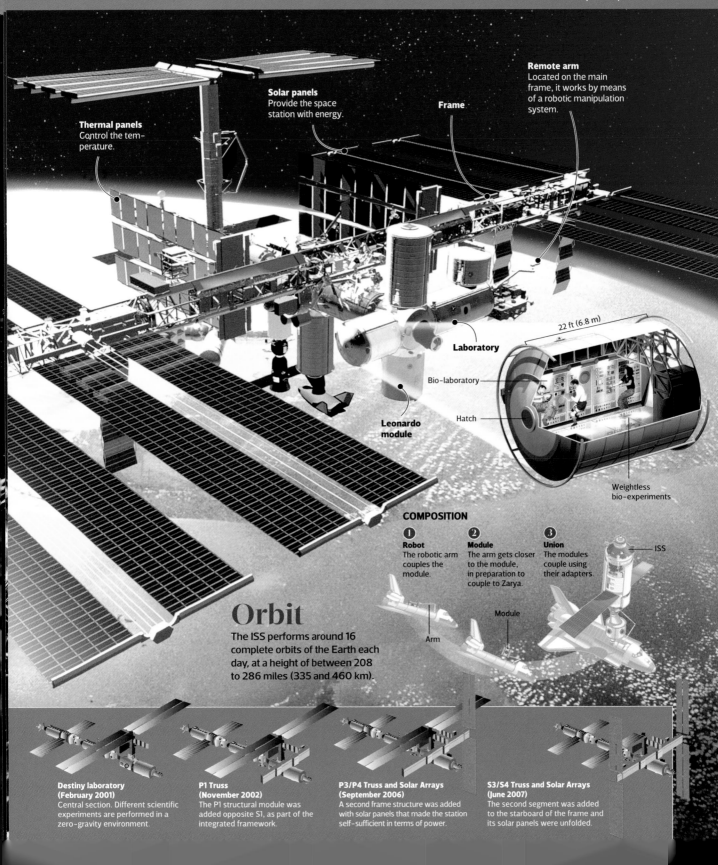

Remote arm
Located on the main frame, it works by means of a robotic manipulation system.

Frame

Solar panels
Provide the space station with energy.

Thermal panels
Control the temperature.

22 ft (6.8 m)

Laboratory

Bio-laboratory

Hatch

Leonardo module

Weightless bio-experiments

COMPOSITION

1 Robot
The robotic arm couples the module.

2 Module
The arm gets closer to the module, in preparation to couple to Zarya.

3 Union
The modules couple using their adapters.

ISS

Module

Arm

Orbit

The ISS performs around 16 complete orbits of the Earth each day, at a height of between 208 to 286 miles (335 and 460 km).

Destiny laboratory (February 2001)
Central section. Different scientific experiments are performed in a zero-gravity environment.

P1 Truss (November 2002)
The P1 structural module was added opposite S1, as part of the integrated framework.

P3/P4 Truss and Solar Arrays (September 2006)
A second frame structure was added with solar panels that made the station self-sufficient in terms of power.

S3/S4 Truss and Solar Arrays (June 2007)
The second segment was added to the starboard of the frame and its solar panels were unfolded.

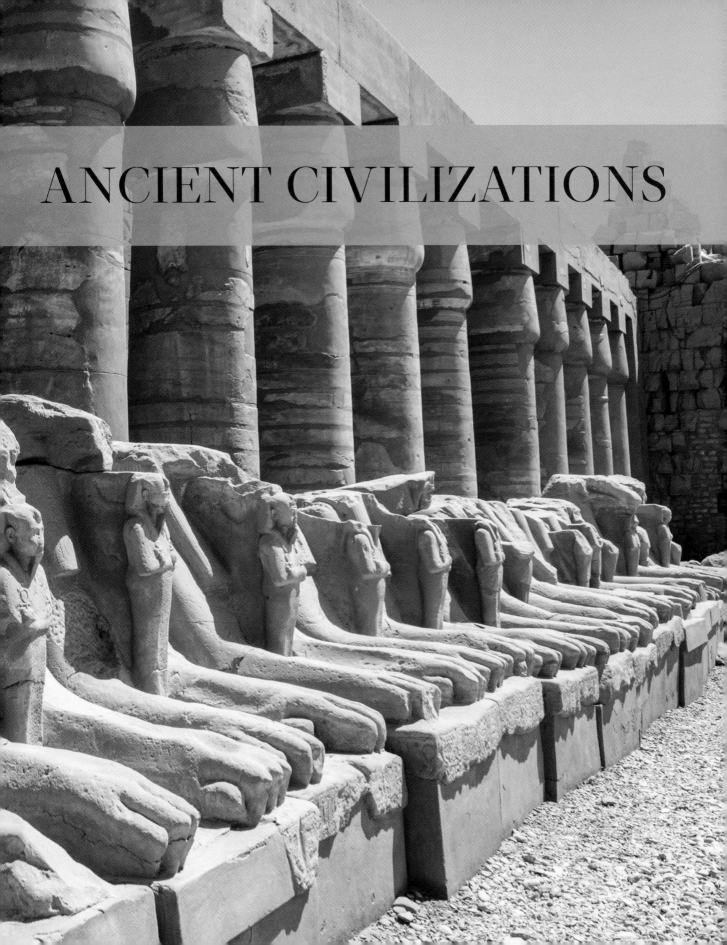

ANCIENT CIVILIZATIONS

Stonehenge

Stonehenge was a temple for the observation of astronomical phenomena. A calendar was used to predict the beginning and end of seasons and to establish order in farming and cattle rearing. An emblem of European architecture in the Bronze Age, this compound represents the main work of an ancient society interested in observing the stars. That started a transition from traditional hunting life to a sedentary lifestyle based on agriculture.

The structure

It is made up of concentric circles of megaliths up to 16 ft (5 m) tall. Perfectly aligned on the ground, they can calculate the trajectory of the Sun and the Moon and indicate solstices and eclipses. ↘

The altar
At the center, a slab of micaceous sand–stone can be found.

First ring
Measuring 98 ft (30 m) in diameter and comprising 30 sandstone monoliths weighing 25 tons each, united by a continuous lintel. Today, only seven pieces remain standing.

Second ring
Formed by smaller blocks of bluestone than those of the outer circle.

FROM THE NEOLITHIC TO THE BRONZE AGE

Every 18.6 years, the moon reaches an extreme azimuth on the horizon, the lunistice. At Stonehenge, the moon lines up with the sun, reflecting the yearning for the age of hunting symbolized by the blue satellite.

----- The course of the sun

----- The course of the moon

Third ring
An assembly of five trilithons arranged in the shape of a horseshoe. Each one is formed by two menhirs crowned by a 14.4 ft (4.4 m) lintel.

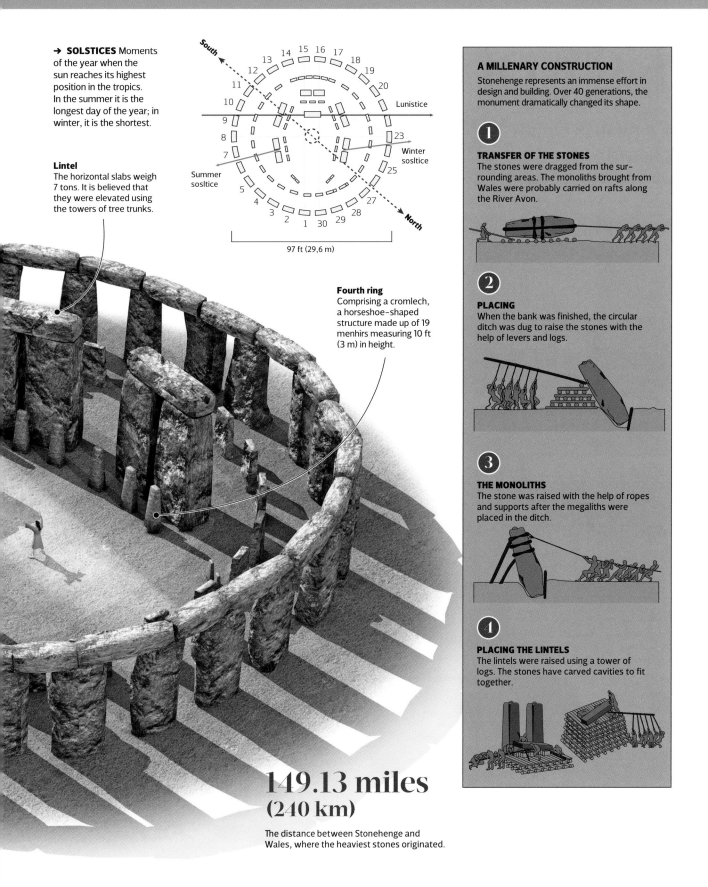

→ **SOLSTICES** Moments of the year when the sun reaches its highest position in the tropics. In the summer it is the longest day of the year; in winter, it is the shortest.

Lintel
The horizontal slabs weigh 7 tons. It is believed that they were elevated using the towers of tree trunks.

South

Lunistice

Winter sosltice

Summer sosltice

North

97 ft (29,6 m)

Fourth ring
Comprising a cromlech, a horseshoe-shaped structure made up of 19 menhirs measuring 10 ft (3 m) in height.

149.13 miles (240 km)

The distance between Stonehenge and Wales, where the heaviest stones originated.

A MILLENARY CONSTRUCTION

Stonehenge represents an immense effort in design and building. Over 40 generations, the monument dramatically changed its shape.

①

TRANSFER OF THE STONES
The stones were dragged from the surrounding areas. The monoliths brought from Wales were probably carried on rafts along the River Avon.

②

PLACING
When the bank was finished, the circular ditch was dug to raise the stones with the help of levers and logs.

③

THE MONOLITHS
The stone was raised with the help of ropes and supports after the megaliths were placed in the ditch.

④

PLACING THE LINTELS
The lintels were raised using a tower of logs. The stones have carved cavities to fit together.

Tutankhamen's tomb

In 1922, British archaeologist Howard Carter found the tomb of Tutankhamen in the Valley of the Kings. This pharaoh had been a member of the XVIII dynasty and ruled between 1333 and 1323 BC. This building is very small to be the last residence of the pharaoh, which means it was not intended to be the young king's tomb. It follows the structure of the other tombs in the valley: a gallery ends in an antechamber that also links with the funerary chamber. Two side chambers have been identified as the "annex" and "the Treasury."

The funerary chamber

The main chamber of the tomb, which contained the pharaoh's coffin, was hidden behind a sealed wall. The entrance was protected by two statues the size of the original Tutankhamen: One represented the young king and the other his "Ka" or spirit.

→

Antechamber
The whole chamber was sealed by walls. When the archaeologist Carter crossed the first door, he found a room full of the pharaoh's objects, many of them were made with gold or carved wood.

The corridor
Both the corridor and the staircases were covered with carved stones, probably extracted from the excavation. There were also valuable pieces on the floor left behind from a robbery.

5.6 ft (1.7 m)

6.5 ft (2 m)

Entrance
Hidden in the rocky soil of the Valley of the Kings, on November 24, 1922, the archaeologist Howard Carter discovered the entrance to Tutankhamen's tomb.

→ THE LOST TOMB It was a matter of luck that Tutankhamen's tomb remained intact. The funerary chamber of the young king was built in the Valley of the Kings; 200 years later the Egyptians dug the sepulchre of the pharaoh Ramses VI. This construction meant that stones covered Tutankhamen's mausoleum.

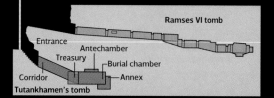

Ramses VI tomb

Entrance
Antechamber
Treasury
Burial chamber
Corridor
Annex
Tutankhamen's tomb

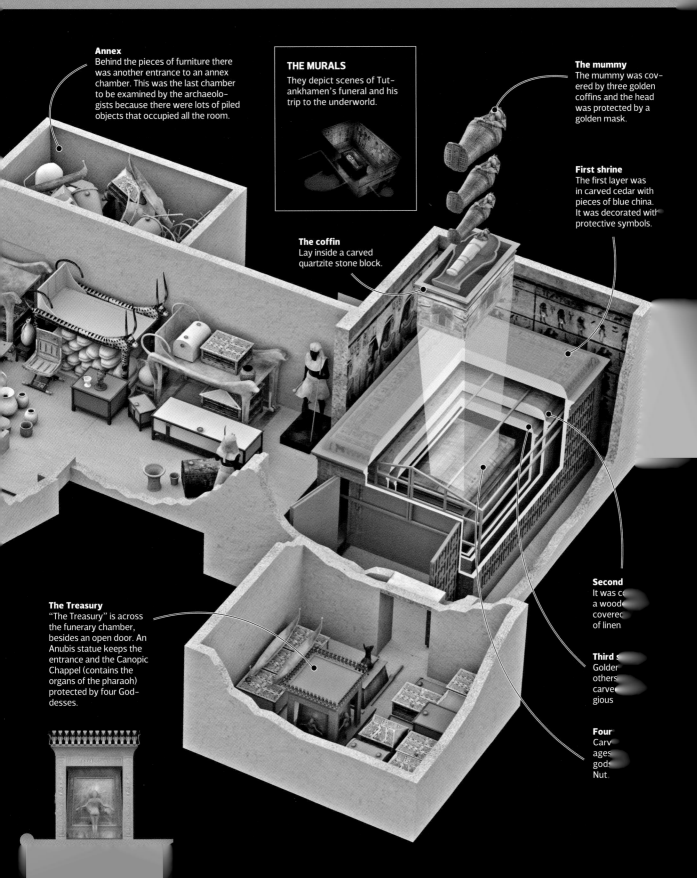

Annex
Behind the pieces of furniture there was another entrance to an annex chamber. This was the last chamber to be examined by the archaeologists because there were lots of piled objects that occupied all the room.

THE MURALS
They depict scenes of Tutankhamen's funeral and his trip to the underworld.

The mummy
The mummy was covered by three golden coffins and the head was protected by a golden mask.

First shrine
The first layer was in carved cedar with pieces of blue china. It was decorated with protective symbols.

The coffin
Lay inside a carved quartzite stone block.

Second
It was c
a wood
covered
of linen

Third s
Golder
others
carve
gious

Four
Carv
ages
gods
Nut.

The Treasury
"The Treasury" is across the funerary chamber, besides an open door. An Anubis statue keeps the entrance and the Canopic Chappel (contains the organs of the pharaoh) protected by four Goddesses.

The Temple of Zeus

Ancient Greeks lived in a world full of gods and heroes. They were immortal, although subject to most human emotions, such as wrath, jealousy, and envy. To worship them, the Greeks had uncountable rites, and as a corollary of their advanced architectural knowledge, they built temples. However, unlike what occurs with modern religions, the Greeks did not worship inside the temples. Rituals, instead, took place outside the temples. The temples are a lasting monument to one of the most advanced cultures of the ancient world.

Zeus sanctuary

Built in the mid-5th century BC on a former famous sanctuary in Olympia, it is considered the perfect model of a Doric-style temple. The excavation and research work that started three centuries ago continues today. ↘

THE THREE STYLES

The first Greek temples date back to the 9th century B.C. Over time, they became more elegant and decorative. They can be divided into three main styles according to their characteristics: Doric, like the Athens Parthenon, Ionian and Corinthian. The beauty of the Greek temples served as a model for the buildings of many later cultures, starting with the Romans, who refined Hellenic architecture.

Pediment
The west pediment depicted the clash between Centaurs, who were guests at the wedding of the Lapith hero Peirithoos, and the Lapithae.

Pillars
Have Doric styling, referring to the ridges around the column.

Stylobate
The temple's ground; a platform that the temple rests upon.

Stereobates
The steps that lead to the temple.

↓ **THE PARTHENON'S FRIEZES** London's British Museum holds around 40% of the Parthenon's huge frieze, considered one of the most important masterpieces of ancient Greece.

ZEUS STATUE
It was one of the Seven Wonders of the Ancient World. Carved in marble by Fidias around 430 BC, it was 39.37 ft (12 m) high. The date and cause of its destruction is unknown.

Elis coins
Minted in the time of Adrian, during the 2nd century, they are the only graphic references of the Zeus statue.

Tiles
Made of marble from Penteliko's mountain.

Opisthodomos (W)
Was like a pronaos, although on the opposite side. It could be connected with the naos. It only existed in temples of great importance.

Naos (N)
Occupied the central space of the temple where the patron deity was located. Only the priests could enter it.

Pronaos (P)
A type of hall that led to the Naos, normally defined by the temple pillars.

Greek Theater

The advent of Greek theater, created in Athens from religious celebrations in honor of the god Dionysus, was foundational for the Western world. The classical masters of tragedy (Aeschylus, Sophocles and Euripides) and comedy (Aristophanes) shaped theatrical plays in the modern sense. In Greek theater, the chorus had a central role in the development of events, plays included songs and dances, a maximum of three actors were permitted in each scene, and women were not allowed to act.

The theatrical architecture

The Greeks established the model of the theatrical architecture whose basic features, improved by the Romans, are still valid today: a building exclusively devoted to plays, whose architectural arrangement clearly separates the space reserved for the public from that reserved for the perfomance. ↘

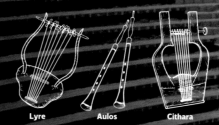

↓ A OPERATIC DRAMA Musical accompaniment from instruments like lyres, flutes, flageolets, and drums made the Greek theater more similar to modern opera than to modern theatre.

Lyre   **Aulos**   **Cithara**

THEATER STRUCTURE
The place where the plays were represented was divided into multiple sections.

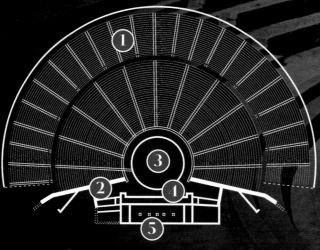

 CAVEA
Part of the theater equipped with bleachers for the public.

 PARODOI
Lateral passageway for the approach to the performing ground.

 ORCHESTRA
Circular space for the chorus, who were members of the cast.

 PROSCENIUM
Platform for actors.

 SKENE
Building for decorative elements and costumes.

300

Tragedies written between Sophocles, Aeschylus, and Euripides.

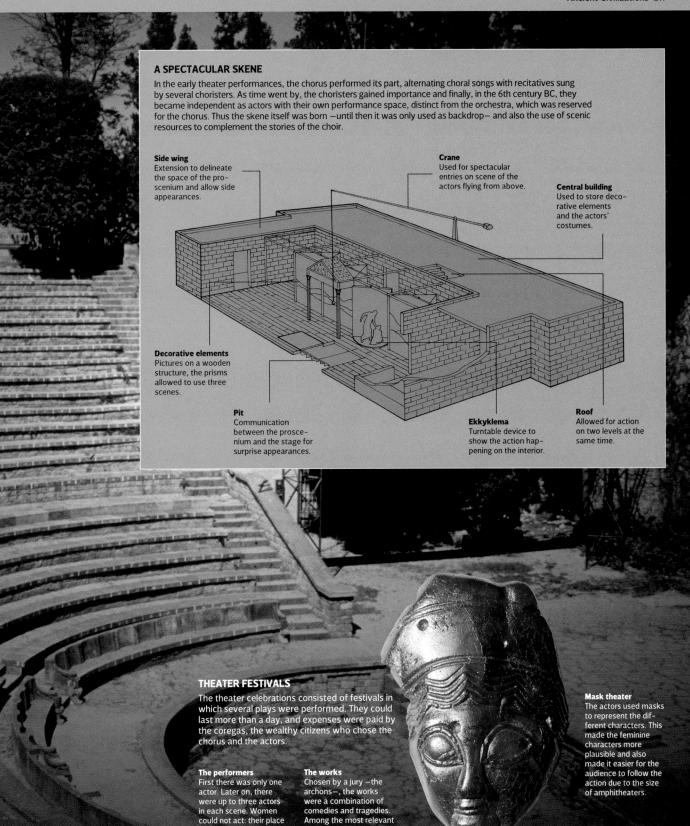

A SPECTACULAR SKENE

In the early theater performances, the chorus performed its part, alternating choral songs with recitatives sung by several choristers. As time went by, the choristers gained importance and finally, in the 6th century BC, they became independent as actors with their own performance space, distinct from the orchestra, which was reserved for the chorus. Thus the skene itself was born —until then it was only used as backdrop— and also the use of scenic resources to complement the stories of the choir.

Side wing
Extension to delineate the space of the pro-scenium and allow side appearances.

Crane
Used for spectacular entries on scene of the actors flying from above.

Central building
Used to store deco-rative elements and the actors' costumes.

Decorative elements
Pictures on a wooden structure, the prisms allowed to use three scenes.

Pit
Communication between the prosce-nium and the stage for surprise appearances.

Ekkyklema
Turntable device to show the action hap-pening on the interior.

Roof
Allowed for action on two levels at the same time.

THEATER FESTIVALS

The theater celebrations consisted of festivals in which several plays were performed. They could last more than a day, and expenses were paid by the coregas, the wealthy citizens who chose the chorus and the actors.

The performers
First there was only one actor. Later on, there were up to three actors in each scene. Women could not act: their place was occupied by men in disguise.

The works
Chosen by a jury —the archons—, the works were a combination of comedies and tragedies. Among the most relevant authors, Aeschylus, Sophocles, Euripides and Aristophanes stand out.

Mask theater
The actors used masks to represent the dif-ferent characters. This made the feminine characters more plausible and also made it easier for the audience to follow the action due to the size of amphitheaters.

Roman aqueducts

The aqueducts were one of the greatest achievements of Roman engineering, and one of the longest–lasting. Rainwater and mountain springs would collect into reservoirs and then aqueducts would bridge the distance between a reservoir and a city. The governing idea behind aqueducts, and the thing that made them successful, was the principle that gravity would carry water along the route.

The structure

The aqueducts of Roman bridges were always made with arches and vaults because it made them more resistant to weathering. Channels could be uncovered or covered with a vaulted roof. ↘

Scaffolding
This wooden structure supported the weight of the arch until the last stone was placed. When it was removed, the stones supported their own weight.

Pieces of pottery
The structure was covered with a mixture of lime and small pieces of crushed pottery to make it waterproof.

LOCATION
To build an aqueduct, the Romans first had to find the right location. The "librator" (the manager of the building) or surveyor looked for traces of vegetation and humidity on the ground.

PILLARS
Even though the height depended on the slope and the irregularities of the ground, the pillars could have a considerable height. They were made of granite and brick.

SCAFFOLDING
They built vertical supports from the ground as construction progressed.

STONES
The Romans excavated many quarries to find the materials to build the aqueduct. If the side of the mountain was sloped, the limestone was covered with layers of concrete and brick.

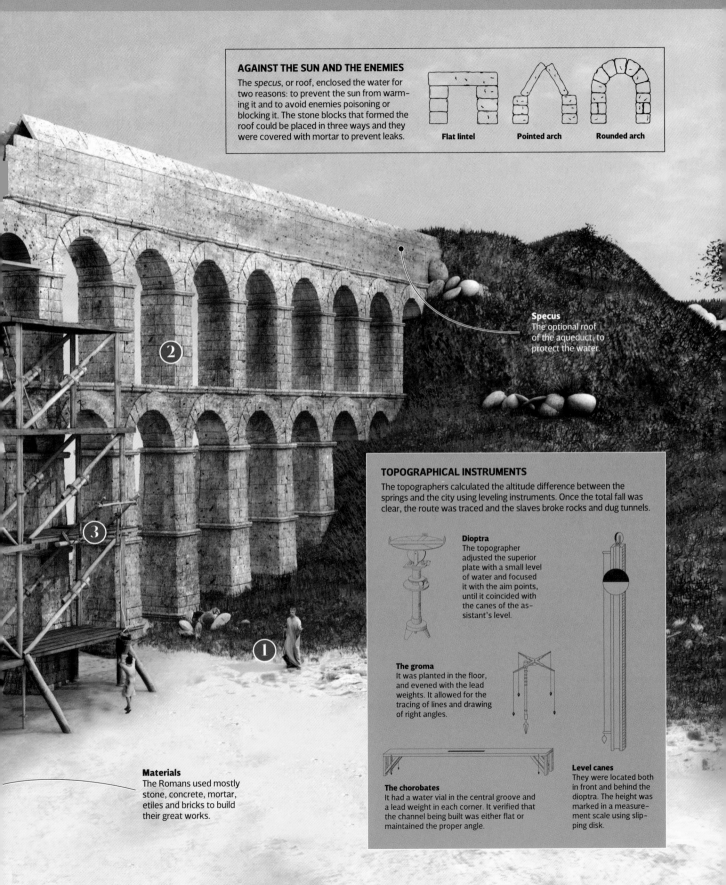

AGAINST THE SUN AND THE ENEMIES

The *specus*, or roof, enclosed the water for two reasons: to prevent the sun from warming it and to avoid enemies poisoning or blocking it. The stone blocks that formed the roof could be placed in three ways and they were covered with mortar to prevent leaks.

Flat lintel　　**Pointed arch**　　**Rounded arch**

Specus
The optional roof of the aqueduct, to protect the water.

TOPOGRAPHICAL INSTRUMENTS

The topographers calculated the altitude difference between the springs and the city using leveling instruments. Once the total fall was clear, the route was traced and the slaves broke rocks and dug tunnels.

Dioptra
The topographer adjusted the superior plate with a small level of water and focused it with the aim points, until it coincided with the canes of the assistant's level.

The groma
It was planted in the floor, and evened with the lead weights. It allowed for the tracing of lines and drawing of right angles.

Level canes
They were located both in front and behind the dioptra. The height was marked in a measurement scale using slipping disk.

The chorobates
It had a water vial in the central groove and a lead weight in each corner. It verified that the channel being built was either flat or maintained the proper angle.

Materials
The Romans used mostly stone, concrete, mortar, etiles and bricks to build their great works.

Machu Picchu

Machu Picchu was built by the middle of the fifteenth century, by the Inca Empire, in modern Peru. It was located among the peaks of Huayna Picchu (young hill) and Machu Picchu (old hill). It was located in a strategic area, protected by two narrow passes and the river Urubamba. It was discovered on July 24, 1911, by, H. Bingham and M. Arteaga. Machu Picchu was declared a UNESCO World Heritage Site in 1983.

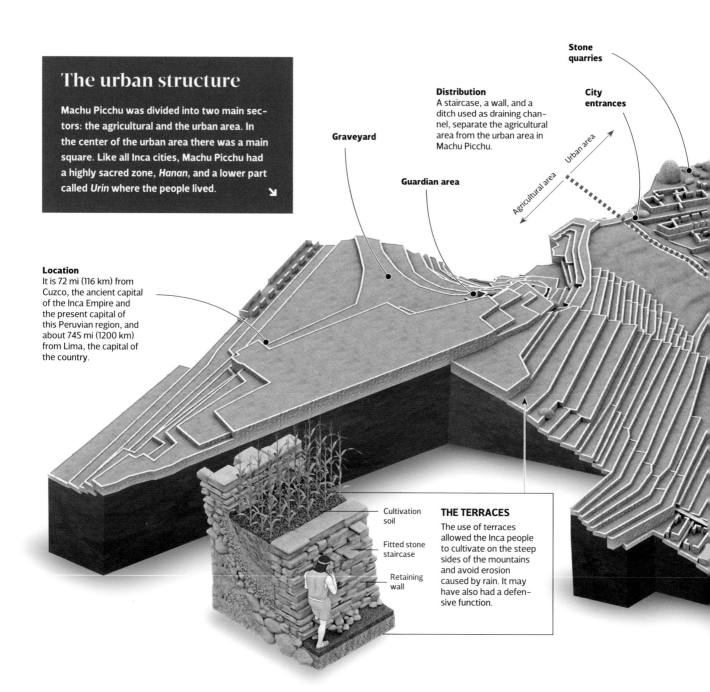

The urban structure

Machu Picchu was divided into two main sectors: the agricultural and the urban area. In the center of the urban area there was a main square. Like all Inca cities, Machu Picchu had a highly sacred zone, *Hanan*, and a lower part called *Urin* where the people lived. ↘

Stone quarries

Distribution
A staircase, a wall, and a ditch used as draining channel, separate the agricultural area from the urban area in Machu Picchu.

City entrances

Graveyard

Urban area

Agricultural area

Guardian area

Location
It is 72 mi (116 km) from Cuzco, the ancient capital of the Inca Empire and the present capital of this Peruvian region, and about 745 mi (1200 km) from Lima, the capital of the country.

Cultivation soil

Fitted stone staircase

Retaining wall

THE TERRACES
The use of terraces allowed the Inca people to cultivate on the steep sides of the mountains and avoid erosion caused by rain. It may have also had a defensive function.

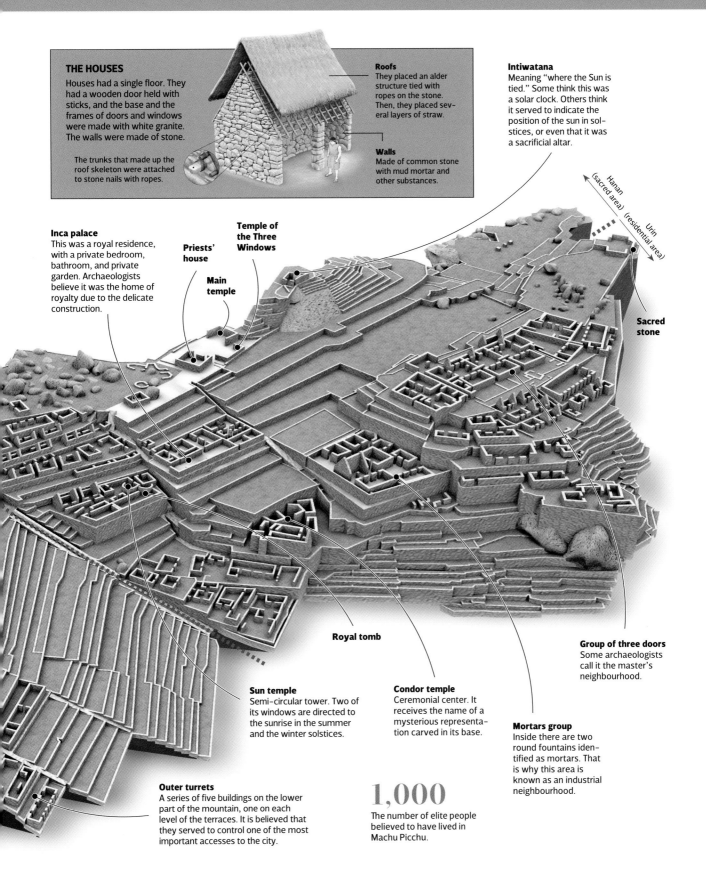

THE HOUSES

Houses had a single floor. They had a wooden door held with sticks, and the base and the frames of doors and windows were made with white granite. The walls were made of stone.

The trunks that made up the roof skeleton were attached to stone nails with ropes.

Roofs
They placed an alder structure tied with ropes on the stone. Then, they placed several layers of straw.

Walls
Made of common stone with mud mortar and other substances.

Intiwatana
Meaning "where the Sun is tied." Some think this was a solar clock. Others think it served to indicate the position of the sun in solstices, or even that it was a sacrificial altar.

Hanan (sacred area)

Urin (residential area)

Sacred stone

Inca palace
This was a royal residence, with a private bedroom, bathroom, and private garden. Archaeologists believe it was the home of royalty due to the delicate construction.

Priests' house

Temple of the Three Windows

Main temple

Royal tomb

Group of three doors
Some archaeologists call it the master's neighbourhood.

Sun temple
Semi-circular tower. Two of its windows are directed to the sunrise in the summer and the winter solstices.

Condor temple
Ceremonial center. It receives the name of a mysterious representation carved in its base.

Mortars group
Inside there are two round fountains identified as mortars. That is why this area is known as an industrial neighbourhood.

Outer turrets
A series of five buildings on the lower part of the mountain, one on each level of the terraces. It is believed that they served to control one of the most important accesses to the city.

1,000
The number of elite people believed to have lived in Machu Picchu.

Viking ship

The Vikings were expert navigators who dominated the maritime routes and rivers of north-east Europe between the eighth and ninth centuries. They also reached certain places on the Mediterranean coast. For raiding and looting they used a long, narrow, and lightweight boat called the *drakkar*, while for commercial activity they used a broader, flatter vessel, especially designed for the transport of timber, wool, hides, wheat, and even slaves.

THE HISTORY OF THE VIKING SHIP

Fishing activity off the coast of Scandinavia fostered the construction of boats. How they are made, which is shown here, is known thanks to different archaeological discoveries of boat remains as well as sketches and reliefs on flat stones.

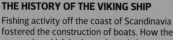

Neolithic canoe, c. 3500 BC

Hjortspring boat, c. 350 BC

Halsnøy boat, c. 100

Nydam boat, c. 350

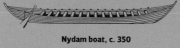

Kvalsund boat, c. 700

The Gokstad ship

The discovery in 1880 of the Gokstad ship in southern Norway advanced our current knowledge of the Vikings. This drakkar dating from around 900, measures more than 75 ft (23 m) in length and, including the rigging, weighs around 20 tons. ↘

Figurehead
Skilled craftsmen, the Vikings carved a symbolic animal into the wood. It was a combination of a dragon and a snake wrapped around itself.

The hull
The planks at the bottom were just 1 in (2.6 cm) thick. The tenth row had to be stronger, 2 in (4.3 cm), as it was on the waterline.

The keel
Made of a single piece of oak over 82 ft (25 m) long, the keel provided great strength and allowed the ship to be navigated in just 3 ft (1 m) of water.

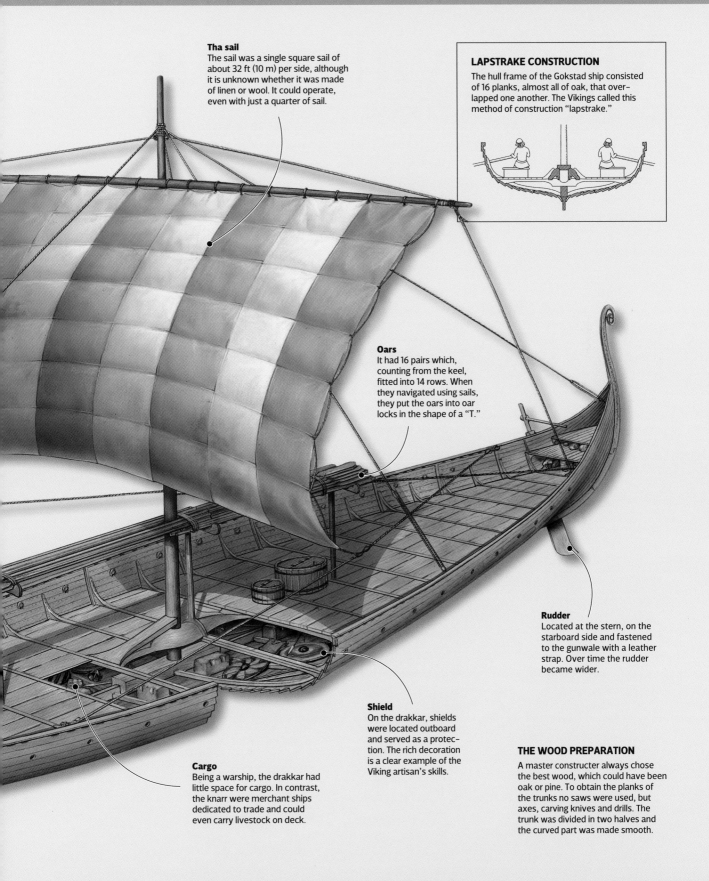

Tha sail
The sail was a single square sail of about 32 ft (10 m) per side, although it is unknown whether it was made of linen or wool. It could operate, even with just a quarter of sail.

LAPSTRAKE CONSTRUCTION
The hull frame of the Gokstad ship consisted of 16 planks, almost all of oak, that overlapped one another. The Vikings called this method of construction "lapstrake."

Oars
It had 16 pairs which, counting from the keel, fitted into 14 rows. When they navigated using sails, they put the oars into oar locks in the shape of a "T."

Rudder
Located at the stern, on the starboard side and fastened to the gunwale with a leather strap. Over time the rudder became wider.

Shield
On the drakkar, shields were located outboard and served as a protection. The rich decoration is a clear example of the Viking artisan's skills.

Cargo
Being a warship, the drakkar had little space for cargo. In contrast, the knarr were merchant ships dedicated to trade and could even carry livestock on deck.

THE WOOD PREPARATION
A master constructer always chose the best wood, which could have been oak or pine. To obtain the planks of the trunks no saws were used, but axes, carving knives and drills. The trunk was divided in two halves and the curved part was made smooth.

Medieval castle

Castles were the fortified residences of kings, noblemen, and lords. They were usually built on high ground as a strategic measure to control and contain external threats and attacks. The castle also served as a refuge for the peasants of the manor in the event of armed aggression and prolonged sieges. From the thirteenth century, the development of offensive weapons by besiegers forced the structure of the interior of the castles to be changed and their defence systems to be improved.

The keep
This was the main tower and served as the residence of the lord and his family. Their riches were stored in the lower section.

Circular towers
The floors were accessed by spiral staircases that descended to the basement. Many towers had their own well in order to avoid depending on the outside.

Drawbridge
Often the castle was surrounded by a moat, to prevent access by the enemy, and could only be entered via a drawbridge.

MERLONS

The separation gaps between the battlements left soldiers on the walkway exposed. This problem was resolved during the thirteenth century with the creation of *merlons*, barriers usually made of wood or metal that could be fixed or removable.

Fixed wooden merlon

Iron, detachable

Wood, detachable

Wall
The entire enclosure was sur-rounded by a high, thick wall. Its towers and battlements had holes from which to shoot at the enemy.

Walkway
Narrow walkway on top of the wall. Allowed sentinels to keep watch over the outside and to organize themselves in the event of an attack.

Parade ground
Central space of the castle from which the rest of the rooms were accessed. It was usually near the chapel and the stables or armory.

Chapel
A small church, located inside the castle walls.

Bread oven
Located within the castle to ensure there was a supply of bread in the event of attack.

Banquets and Parties

During peace times, kings and nobles organized large parties that could last for several days. Banquets were complemented with tourna-ments, duels, and literary competitions, in which minstrels and troubadours participated.

MACHINES OF WAR

Medieval siege machinery

Until the arrival of gunpowder, the siege engines of the Middle Ages were, in essence, an evolution of those inherited from ancient times: ballistae, catapults, battering rams and siege towers. But the most significant innovation of this period was the trebuchet, introduced to the West around the twelfth century, which had a longer range than a catapult and was capable of demolishing walls.

SIEGE TOWER

This could transport and protect many men on their approach to an enemy's walls. Once there, they opened a breach in the walls or doors, or a drawbridge was lowered allowing them to enter from above.

26.25 ft (8 m)

Drawbridge
Once at the wall, this was lowered, allowing the soldiers to enter.

Structure
Square-based tower with three stories, connected by ladders, and protected by wooden siding and shields.

Battering ram
A log with an iron head, used to break down doors in fortifications by repeatedly striking them.

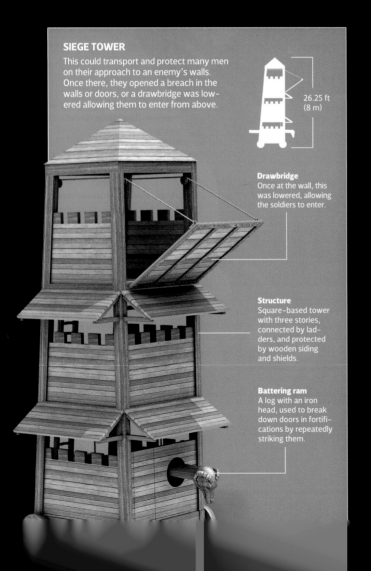

Trebuchet

This was used to launch enormous projectiles at the walls or structures of a fortification to destroy them. Similar to a large catapult, it used counterweights and gravity instead of torsion. It had a range of about 650 ft (200 m).
→

Ammunition
Large rocks and stones were the main ammunition, but there is evidence of other materials like dung and dead animals. The latter were hurled into the besieged city to spread disease.

Levers
When they were operated, the arm descended and the counterweight rose.

Stakes
These secured the structure to the ground to prevent it from lifting when fired.

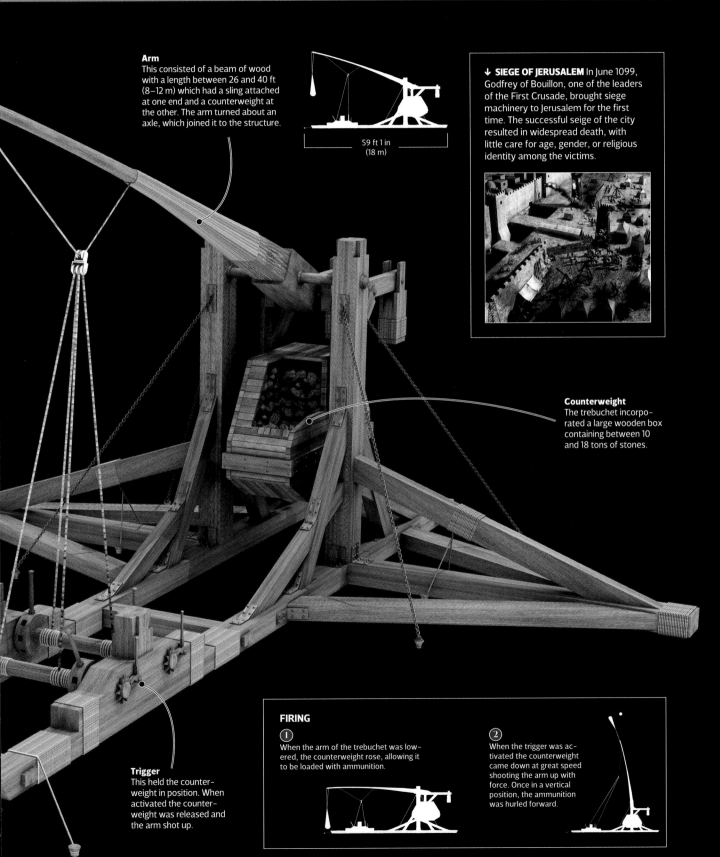

Arm
This consisted of a beam of wood with a length between 26 and 40 ft (8–12 m) which had a sling attached at one end and a counterweight at the other. The arm turned about an axle, which joined it to the structure.

59 ft 1 in
(18 m)

↓ SIEGE OF JERUSALEM In June 1099, Godfrey of Bouillon, one of the leaders of the First Crusade, brought siege machinery to Jerusalem for the first time. The successful seige of the city resulted in widespread death, with little care for age, gender, or religious identity among the victims.

Counterweight
The trebuchet incorporated a large wooden box containing between 10 and 18 tons of stones.

Trigger
This held the counterweight in position. When activated the counterweight was released and the arm shot up.

FIRING

① When the arm of the trebuchet was lowered, the counterweight rose, allowing it to be loaded with ammunition.

② When the trigger was activated the counterweight came down at great speed shooting the arm up with force. Once in a vertical position, the ammunition was hurled forward.

Korean Turtle Ships

In the late sixteenth century the Japanese attempted to conquer the Korean Peninsula as a prelude to the invasion of China. However, they encountered fierce resistance and an efficient fleet of "turtle boats," the main deck of which was protected by armor plating, possibly metal, with large "pins" protruding from it. Although this cover was not useful for defending the ship from enemy arrows and cannon, its main function was to prevent the ship from being boarded.

Eastern sea fortress

The "turtle ships" or Geobukseon were strong and stable, but also light, fast, and highly maneuverable ships. For some historians, these are considered the first battleships in history.

DIMENSIONS

131 ft (40 m)

23 ft (7 m)

Although there is no absolute certainty, it is likely they measured just under 131 ft (40 m) in length.

YI SUN-SIN
Korean admiral (1545–1598), famous for commissioning the construction in 1592 several "turtle ship" from some old blueprints. Using a small fleet, he defended his country's from repeated conquest attempts of the Japanese. He died from wounds during a battle against the Japanese enemy.

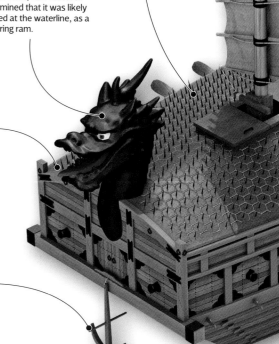

Sails
There were two sails reinforced with numerous rods that gave them great strength. In combat, masts and sails could be removed.

Spikes
The shell was covered with metal spikes to prevent boarding actions.

Dragon-shaped head
Although according to the drawings of the eighteenth century, it was situated as a figurehead, new research determined that it was likely located at the waterline, as a battering ram.

Shell
Made of hexagonal plates, it covered the main deck. It has not been determined whether it was made of iron or wood.

Anchor
Made of wood and large in size, it was another distinctive feature of this type of ship.

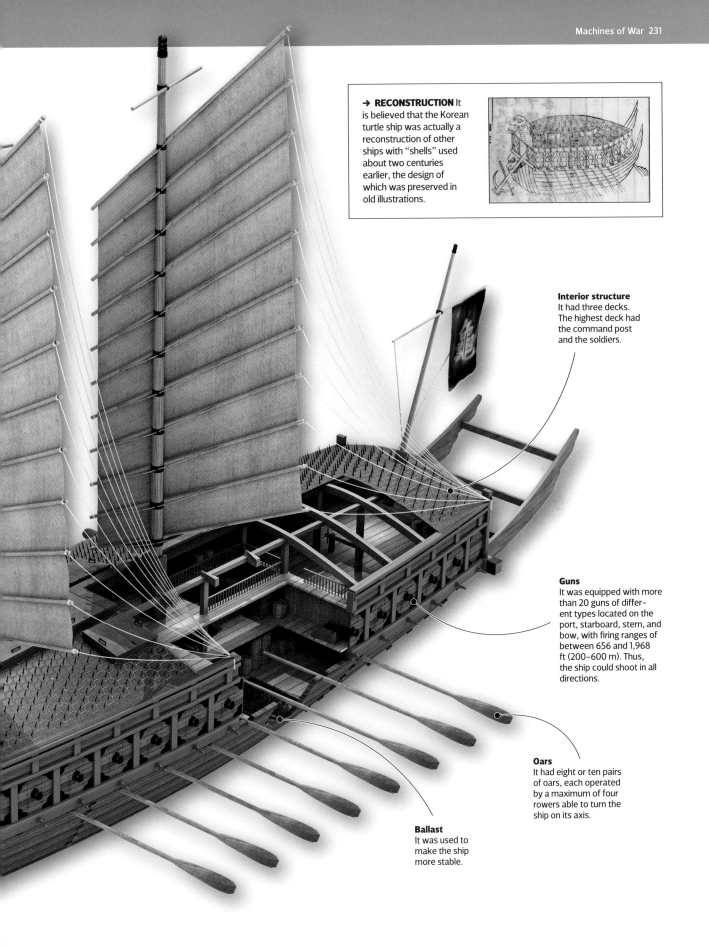

→ RECONSTRUCTION It is believed that the Korean turtle ship was actually a reconstruction of other ships with "shells" used about two centuries earlier, the design of which was preserved in old illustrations.

Interior structure
It had three decks. The highest deck had the command post and the soldiers.

Guns
It was equipped with more than 20 guns of different types located on the port, starboard, stern, and bow, with firing ranges of between 656 and 1,968 ft (200–600 m). Thus, the ship could shoot in all directions.

Oars
It had eight or ten pairs of oars, each operated by a maximum of four rowers able to turn the ship on its axis.

Ballast
It was used to make the ship more stable.

Battleship Mikasa

At the end of the nineteenth century, Japan embarked upon a rapid industrialization and economic development process which would soon see the country become a world power. Such power was demonstrated during the 1904–1905 Russo–Japanese War, sparked by Russian expansion into East Asia and its desire to conquer Korea. The stunning Japanese victory underscored the superiority of its navy, the backbone of which were battleships and heavily armored warships.

Flagship

The *Mikasa*, Admiral Togo's flagship, was decisive in the Japanese victories in the attack on Port Arthur (1904) and the naval Battle of Tsushima (1905). After exploding due to a short circuit, it was rebuilt in 1922 as a national monument at the Military Port of Yokosuka in Japan. ⤵

Command bridge
Reinforced with steel armor plating, it was the nerve center of the battleship. Officers would meet here to make decisions and lead a crew of 850 people.

Artillery
Its main cannons aside, the *Mikasa* was equipped with a further 46 cannons to increase its firepower: 14 at 154 mm caliber, 20 at 76 mm, 8 at 47 mm and 4 at 35 mm.

Main guns
The *Mikasa* was equipped with two double steel guns, one aft and one on the bow, which were able to shoot three 305 mm shells every two minutes, with a range of 6.21 mi (10 km). The forward turret had 40 men assigned.

Emblem
The bow sported the Japanese imperial emblem, a golden chrysanthemum flower.

DIMENSIONS

433 ft (132 m)

73.16 ft (22.3 m)

Lifeboat
Strategically located in case a torpedo should cause a leak and lead to a slow-sinking process. However, the devices were rendered useless if an explosion occurred in the magazine or in the boilers.

THE JAPANESE FLEET

Japan entrusted the construction of its fleet to Britain, France and Italy, all interested in curbing Russian power in Asia. The first Japanese battleship, the *Hiei*, was launched in 1877 from British shipyards, just like the *Mikasa* (1900). The fleet also included captured Chinese battleships.

Telegraph cabin
Incorporated into the Navy in the early twentieth century, wireless telegraphy was decisive for the development of Japanese tactics in the Battle of Tsushima.

H. Togo

Admiral Heihachiro Togo (1848–1934) was the architect of the Japanese Imperial Navy. His strategy was decisive at the Battle of Tsushima (May 27, 1905).

THE IRONCLAD

They were the first modern battleships: the first steam warships armored with iron or steel plates that replaced the wooden vessels, which were very vulnerable to explosive shells. The French Navy was the first to build an ironclad, named *La Gloire* (1859). The British answered a year later with the manufacturing of *HMS Warrior*.

La Gloire
Illustration of the French warship, the first ironclad in history.

Propulsion
The steam propulsion system gave the battleships an advantage over sail boats during the Crimean War (1854–1855). The fumes and steam generated during combustion were released through large chimneys.

Storage
Carbon deposits accumulated fuel for 25 boilers and two engines.

Armored
The steel shield protected the hull, and especially the bridge.

First World War artillery

Trench warfare put the basic principle of artillery to the test: firing on the enemy from further and further away and with greater power. With huge guns and howitzers, heavy artillery would be essential on the Western Front to achieve victory in numerous battles, since only these weapons had the range, caliber, and power needed to destroy the fortified positions of the enemy lines.

"Big Bertha"

German siege howitzer, manufactured by Krupp and used both on the Western and Eastern fronts. Weighing in at 42 tons, it reduced the Belgian forts of Namur and Liège to rubble. ↘

DIMENSIONS

19.3 ft (15,88 m)

Loading
It used the sliding–wedge breech mechanism.

Barrel
16.5 in (420 mm) in diameter, it fired 1,830 lb (830 kg) shells a distance of around 5.8 mi (9.3 km) at a speed of 1,300 ft/sec (400 m/second).

Logistics and operation
Over 200 people were needed to move, set up, and operate this howitzer. Its assembly took six hours and it required 12 men just to operate it.

Mobility
It had very limited mobility and was transported in pieces on four trailers towed by two Daimler Benz tractors.

Aiming arc
It was fitted with a wedge, which was buried in the ground to anchor the gun, and a guide used to direct it.

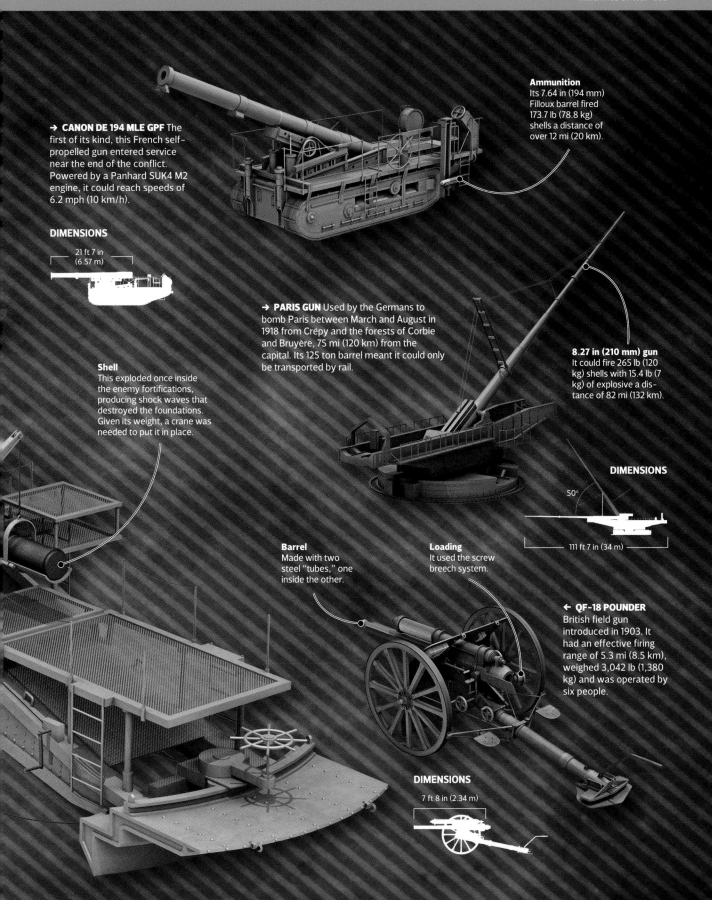

Ammunition
Its 7.64 in (194 mm) Filloux barrel fired 173.7 lb (78.8 kg) shells a distance of over 12 mi (20 km).

→ **CANON DE 194 MLE GPF** The first of its kind, this French self-propelled gun entered service near the end of the conflict. Powered by a Panhard SUK4 M2 engine, it could reach speeds of 6.2 mph (10 km/h).

DIMENSIONS

21 ft 7 in (6.57 m)

→ **PARIS GUN** Used by the Germans to bomb Paris between March and August in 1918 from Crépy and the forests of Corbie and Bruyère, 75 mi (120 km) from the capital. Its 125 ton barrel meant it could only be transported by rail.

8.27 in (210 mm) gun
It could fire 265 lb (120 kg) shells with 15.4 lb (7 kg) of explosive a distance of 82 mi (132 km).

Shell
This exploded once inside the enemy fortifications, producing shock waves that destroyed the foundations. Given its weight, a crane was needed to put it in place.

DIMENSIONS

50°

111 ft 7 in (34 m)

Barrel
Made with two steel "tubes," one inside the other.

Loading
It used the screw breech system.

← **QF-18 POUNDER**
British field gun introduced in 1903. It had an effective firing range of 5.3 mi (8.5 km), weighed 3,042 lb (1,380 kg) and was operated by six people.

DIMENSIONS

7 ft 8 in (2.34 m)

Fokker Dr.I

Using a Sopwith triplane captured from the British, the Germans developed their own version: the Fokker Dr.I. The first two were delivered in August 1917 to the squadron of Manfred von Richthofen (known to the Allies as the "Red Baron"), who took one as his personal plane. The other Fokker went to Werner Voss who shot down twenty planes in 24 days. Its exceptional maneuverability made it ideal for the kind of air combat —dog fighting— which was being waged at the time on the Western Front.

Cockpit

To balance the heavy engine the cockpit had to be set back with respect to the initial prototype. The pilot had excellent vision during flight, although not for takeoff. ↘

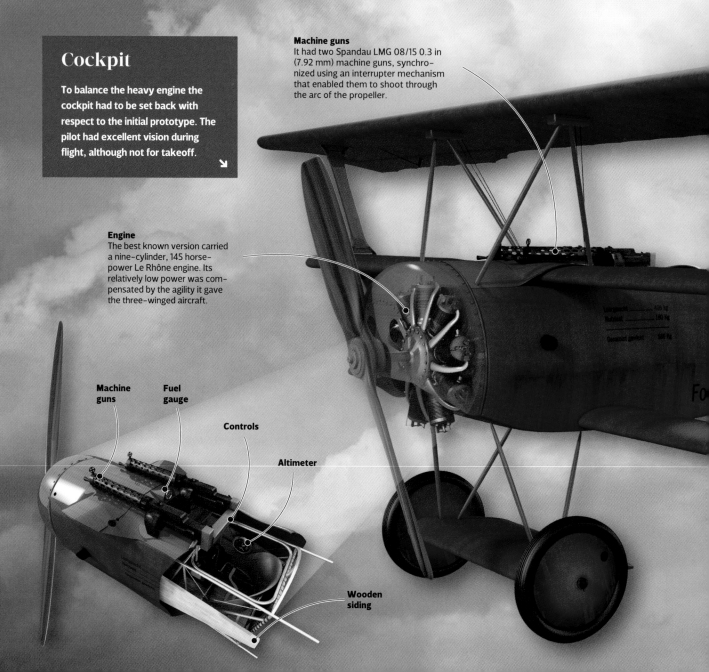

Machine guns
It had two Spandau LMG 08/15 0.3 in (7.92 mm) machine guns, synchronized using an interrupter mechanism that enabled them to shoot through the arc of the propeller.

Engine
The best known version carried a nine-cylinder, 145 horsepower Le Rhône engine. Its relatively low power was compensated by the agility it gave the three-winged aircraft.

Machine guns

Fuel gauge

Controls

Altimeter

Wooden siding

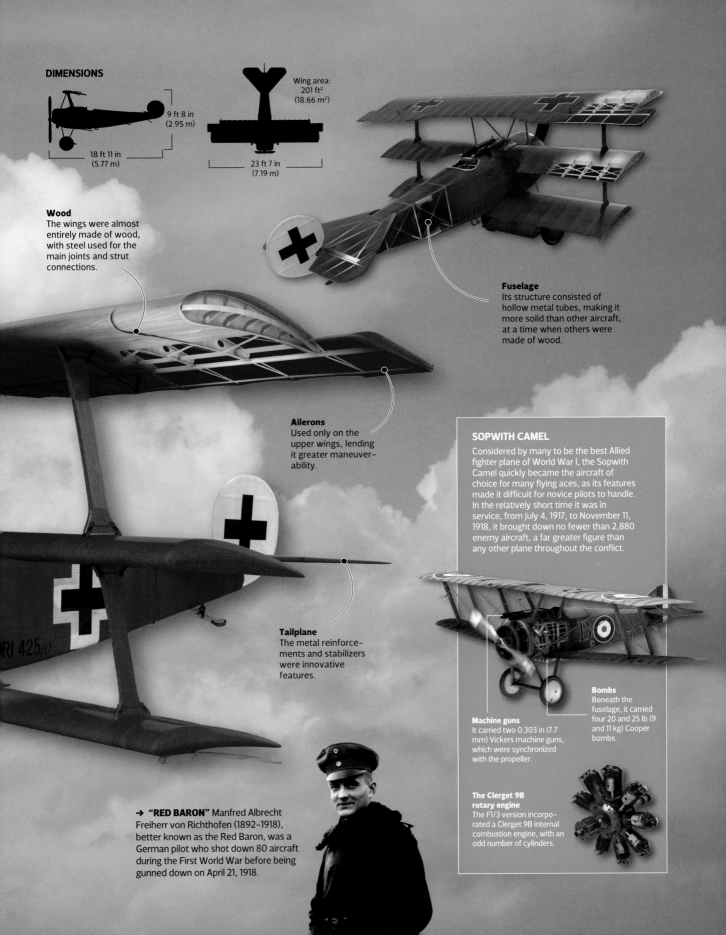

DIMENSIONS

9 ft 8 in
(2.95 m)

18 ft 11 in
(5.77 m)

Wing area:
201 ft²
(18.66 m²)

23 ft 7 in
(7.19 m)

Wood
The wings were almost entirely made of wood, with steel used for the main joints and strut connections.

Fuselage
Its structure consisted of hollow metal tubes, making it more solid than other aircraft, at a time when others were made of wood.

Ailerons
Used only on the upper wings, lending it greater maneuver-ability.

Tailplane
The metal reinforce-ments and stabilizers were innovative features.

SOPWITH CAMEL

Considered by many to be the best Allied fighter plane of World War I, the Sopwith Camel quickly became the aircraft of choice for many flying aces, as its features made it difficult for novice pilots to handle. In the relatively short time it was in service, from July 4, 1917, to November 11, 1918, it brought down no fewer than 2,880 enemy aircraft, a far greater figure than any other plane throughout the conflict.

Machine guns
It carried two 0.303 in (7.7 mm) Vickers machine guns, which were synchronized with the propeller.

Bombs
Beneath the fuselage, it carried four 20 and 25 lb (9 and 11 kg) Cooper bombs.

The Clerget 9B rotary engine
The F1/3 version incorpo-rated a Clerget 9B internal combustion engine, with an odd number of cylinders.

→ **"RED BARON"** Manfred Albrecht Freiherr von Richthofen (1892–1918), better known as the Red Baron, was a German pilot who shot down 80 aircraft during the First World War before being gunned down on April 21, 1918.

T-34 tank

The former Soviet Union produced some of the most envied and copied light and heavy tanks in history. The most emblematic, the armoured T-34, was crucial for halting the progress of Hitler's troops on the Eastern Front during World War II, tipping the balance of the war in favour of the Allies. In recent decades, the T-72, exported to various countries, stands as Russia's second most manufactured armoured vehicle.

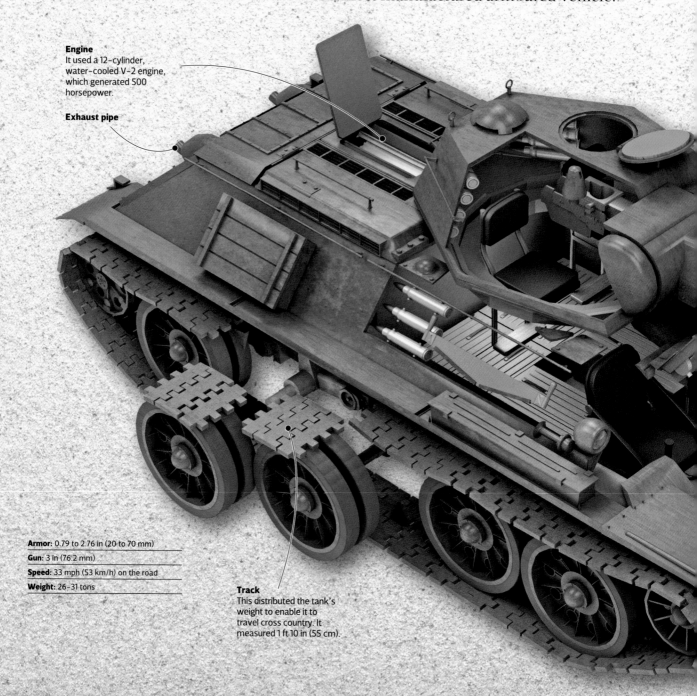

Engine
It used a 12-cylinder, water-cooled V-2 engine, which generated 500 horsepower.

Exhaust pipe

Armor: 0.79 to 2.76 in (20 to 70 mm)

Gun: 3 in (76.2 mm)

Speed: 33 mph (53 km/h) on the road

Weight: 26-31 tons

Track
This distributed the tank's weight to enable it to travel cross country. It measured 1 ft 10 in (55 cm).

The most effective

The success of this Soviet armored vehicle, of which almost 40,000 were produced between 1940 and 1944, is owed to its combination of speed and strength and the fact that it was easy to manufacture.

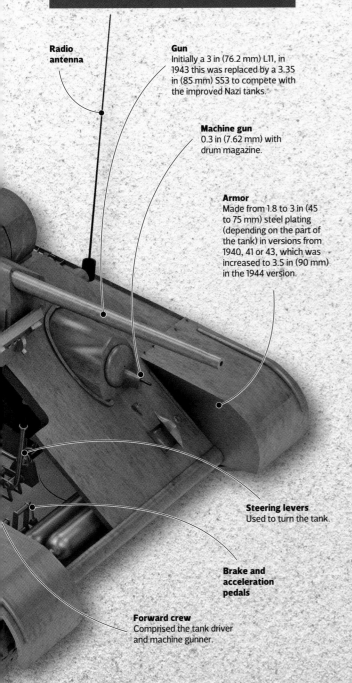

Radio antenna

Gun
Initially a 3 in (76.2 mm) L11, in 1943 this was replaced by a 3.35 in (85 mm) S53 to compete with the improved Nazi tanks.

Machine gun
0.3 in (7.62 mm) with drum magazine.

Armor
Made from 1.8 to 3 in (45 to 75 mm) steel plating (depending on the part of the tank) in versions from 1940, 41 or 43, which was increased to 3.5 in (90 mm) in the 1944 version.

Steering levers
Used to turn the tank.

Brake and acceleration pedals

Forward crew
Comprised the tank driver and machine gunner.

OTHER RED ARMY TANKS

①

T-18
Although it was considered a failure, the T-18 represented the Soviet Union's first steps in tank manufacturing. It was produced between 1928 and 1931.

Armor: 0.63 in (16 mm)	
Gun: 1.46 in (37 mm)	
Speed: 11 mph (17 km/h)	
Weight: 5.9 tons	

②

KV-1
This heavy tank's thick armor rendered it almost indestructible by the German tanks. It entered service in 1940.

Armor: 3.54 in (90 mm)	
Gun: 3 in (76.2 mm)	
Speed: 22 mph (35 km/h)	
Weight: 43 tons	

③

KV-2
Heavy tank, designed to attack fortifications. Produced between 1940 and 1942, it fired powerful shells.

Armor: 4.33 in (110 mm)	
Gun: 5.98 in (152 mm)	
Speed: 22 mph (35 km/h)	
Weight: 58 tons	

④

T-72A
One of the most popular Soviet armored vehicles today in Russia and other parts of the world. It entered service in 1971 and has been exported to countries like Iraq and the former Yugoslavia.

Armor: 3.94 in (100 mm)	
Gun: 4.92 in (125 mm)	
Speed: 37 mph (60 km/h)	
Weight: 41.5 to 44.5 tons	

Junkers Ju-87 Stuka

Their strange and threatening appearance, together with the peculiar howling of their engines, made the German Stuka a fearsome presence in the Polish and French skies during the first two years of World War II. The world had already witnessed their power during the Spanish Civil War, when the A and B models were used by the Condor Legion. The Stuka was able to attack even small and moving targets with great precision.

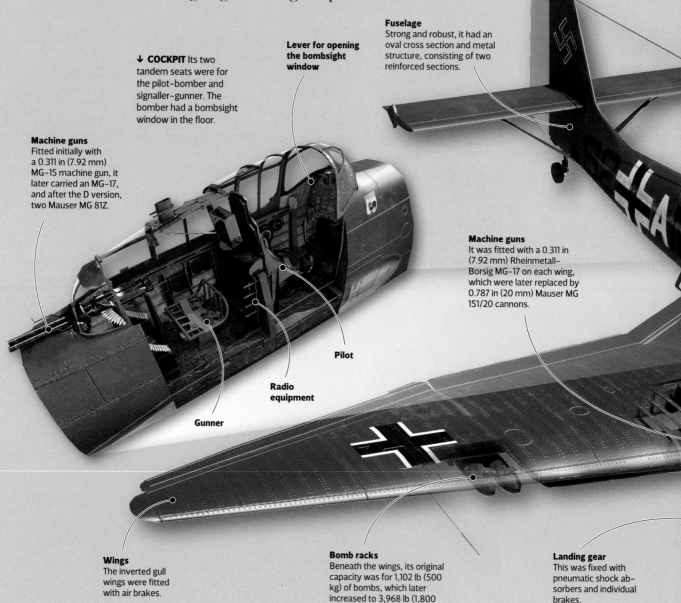

↓ COCKPIT Its two tandem seats were for the pilot-bomber and signaller-gunner. The bomber had a bombsight window in the floor.

Lever for opening the bombsight window

Fuselage
Strong and robust, it had an oval cross section and metal structure, consisting of two reinforced sections.

Machine guns
Fitted initially with a 0.311 in (7.92 mm) MG-15 machine gun, it later carried an MG-17, and after the D version, two Mauser MG 81Z.

Machine guns
It was fitted with a 0.311 in (7.92 mm) Rheinmetall-Borsig MG-17 on each wing, which were later replaced by 0.787 in (20 mm) Mauser MG 151/20 cannons.

Pilot

Radio equipment

Gunner

Wings
The inverted gull wings were fitted with air brakes.

Bomb racks
Beneath the wings, its original capacity was for 1,102 lb (500 kg) of bombs, which later increased to 3,968 lb (1,800 kg) in the D and G series.

Landing gear
This was fixed with pneumatic shock absorbers and individual brakes.

Models

The main version was the JU-87D. Among its variants there was a Navy model, with folding wings for the aircraft carrier Graf Zeppelin; other variants had extra armor plating, torpedoes, anti-tank capabilities, and even one with extended-range.

↓

DIVE-BOMBERS

The height of development for this class of bomber was during World War II, where they were widely used as close support for land troops and the navy. They were designed to be launched at great speed toward a target and drop bombs with great precision, exposing themselves briefly to anti-aircraft fire.

① Sighting the target
The pilot sights the target and sets the automatic pull-out system to the desired altitude.

② Dive
The aircraft dives at an angle of between 65° and 90° at over 311 mph (500 km/h). The dive brakes are opened.

③ Attack
At an altitude of about 1,640 ft (500 m), the pilot launches the bomb or fires the machine gun. At the same time, the automatic system initiates the pull-out.

Fuel
On the wings were two main tanks, each with a capacity of 106 gal (481 l), and another two 66 gal (300 l) tanks.

DIMENSIONS

12 ft 10 in (3.9 m)

37 ft 9 in (11.5 m)

49 ft 3 in (15 m)

Wing area: 362.64 ft² (33.69 m²)

Camouflage
This model was used in the French campaign in 1940 and the Balkans campaign in 1941.

Propeller
This was a VS 5 or VS 11, made from plywood. It had three dual-position blades.

Engine
The engine was held to the fuselage by a cantilevered section formed by two high-strength metal beams.

Ventral pylon
A Siemens ETC 500/A for a SC 500 1,102 lb (500 kg) bomb tilted forwards and launched the bomb in the desired direction.

Unmanned Aerial Vehicles (UAV)

These state-of-the-art, unmanned aircraft are controlled by a ground crew located in military bases miles from the dangers of combat. They are capable of carrying out reconnaissance, attack, and support missions in the harshest of scenarios. Popularly known as "drones," the UAV came into military use in the 1990s.

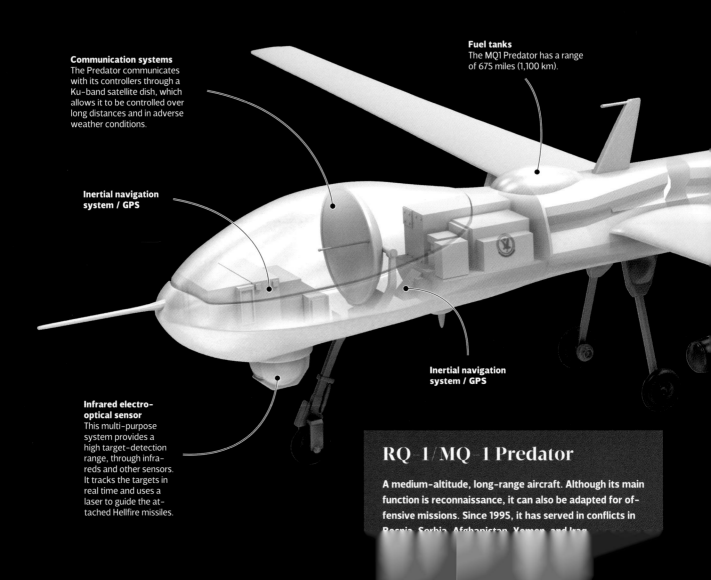

Communication systems
The Predator communicates with its controllers through a Ku-band satellite dish, which allows it to be controlled over long distances and in adverse weather conditions.

Fuel tanks
The MQ1 Predator has a range of 675 miles (1,100 km).

Inertial navigation system / GPS

Inertial navigation system / GPS

Infrared electro-optical sensor
This multi-purpose system provides a high target-detection range, through infra-reds and other sensors. It tracks the targets in real time and uses a laser to guide the attached Hellfire missiles.

RQ-1/MQ-1 Predator

A medium-altitude, long-range aircraft. Although its main function is reconnaissance, it can also be adapted for offensive missions. Since 1995, it has served in conflicts in Bosnia, Serbia, Afghanistan, Yemen, and Iraq.

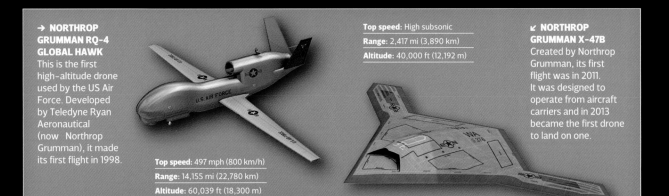

→ NORTHROP GRUMMAN RQ-4 GLOBAL HAWK
This is the first high-altitude drone used by the US Air Force. Developed by Teledyne Ryan Aeronautical (now Northrop Grumman), it made its first flight in 1998.

Top speed: High subsonic
Range: 2,417 mi (3,890 km)
Altitude: 40,000 ft (12,192 m)

Top speed: 497 mph (800 km/h)
Range: 14,155 mi (22,780 km)
Altitude: 60,039 ft (18,300 m)

↙ NORTHROP GRUMMAN X-47B
Created by Northrop Grumman, its first flight was in 2011. It was designed to operate from aircraft carriers and in 2013 became the first drone to land on one.

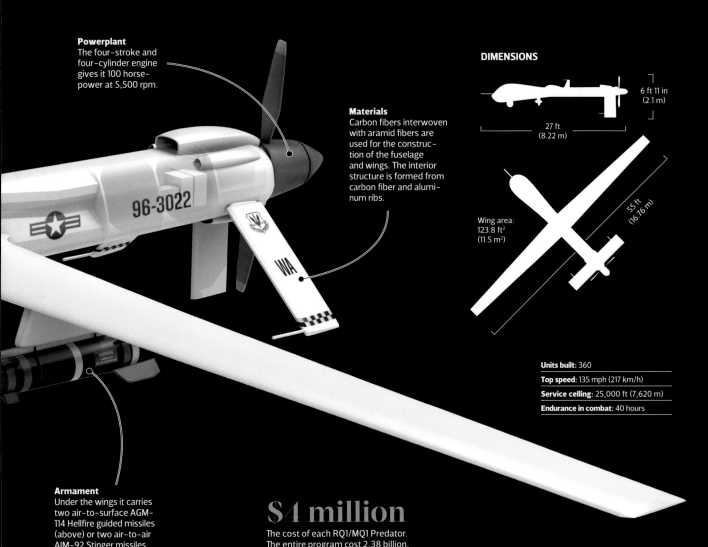

Powerplant
The four-stroke and four-cylinder engine gives it 100 horsepower at 5,500 rpm.

Materials
Carbon fibers interwoven with aramid fibers are used for the construction of the fuselage and wings. The interior structure is formed from carbon fiber and aluminum ribs.

DIMENSIONS

6 ft 11 in (2.1 m)

27 ft (8.22 m)

Wing area: 123.8 ft² (11.5 m²)

55 ft (16.76 m)

Units built: 360
Top speed: 135 mph (217 km/h)
Service ceiling: 25,000 ft (7,620 m)
Endurance in combat: 40 hours

Armament
Under the wings it carries two air-to-surface AGM-114 Hellfire guided missiles (above) or two air-to-air AIM-92 Stinger missiles.

$4 million

The cost of each RQ1/MQ1 Predator. The entire program cost 2.38 billion.

HIMARS Rocket Launchers

The arms industry evolves to respond to current warfare, as has been the case in regions of conflict such as Iraq or Afghanistan. An example of this is the US Army's HIMARS (High Mobility Artillery Rocket System) rocket launcher: a light, easily transportable vehicle capable of hitting its target from a great distance with maximum precision and efficiency. Since their deployment in 2005, these sophisticated trucks equipped with rocket launchers have fired 2,500 projectiles.

Armored cabin
Only three people —the driver, the gunner, and the section chief— travel in this cabin, protected against small weapons and gases.

Deadly weapon

HIMARS was developed in the United States by Lockheed Martin (missile launcher) and BAE Systems (vehicle). It is a device designed to attack artillery positions, anti-aircraft defences, military or logistics vehicles or concentrations of troops from a given distance (up to 186 miles/300 km), with the advantage of being able to abandon its firing position quickly, before being located and counter-attacked. →

PAD
HIMARS can launch six guided missiles (GMLRS) to a distance of up to 43 miles (70 km), or a tactical ATACMS missile, with a range of up to 186 miles (300 km). Its level of accuracy is excellent, thanks to the GPS system incorporated on the projectiles.

Six in less than a minute
The launch sequence for a six GMLRS-missile load is around 45 seconds.

Compatibility
The launch pad can be used for different 8.93 in (227 mm) MLRS-type missile models.

Launcher
Employs a hydraulic lift system. It takes around 15 minutes to prepare the shot, but less than 20 seconds to take aim.

Light
The truck weighs just 10,000 lb (5 tons), meaning it can be transported on a Hercules C-130, one of the most common US Air Force aircraft.

SATELLITE-GUIDED MISSILES

The US Army used the GPS satellite navigation system for the first time in the Gulf War, from 1990–91, to locate and position its troops. At the same time, its Russian counterpart, the GLONASS, was under development. Today, both systems are used for missile and bomb guidance, among other functions.

→ **M31 MISSILES** The first M26 projectile models used by HIMARS, with a range of just 20 mi (32 km), would eventually give way to the new GMLRS M31 guided missiles. One of the main advantages of this extremely accurate weapon, with just a 33 ft (10 m) margin of error, is that it reduces collateral damage.

GPS
Guide and control device, with GPS antenna.

Explosive core
Each M31 contains around 198 lbs (90 kg) of controlled fragmentation explosives.

Multifuse
Equipped with three different adjustments, depending on the target: bunkers, open-air target, etc.

GMLRS motor

6X6 drive
Unlike tracked vehicles, on which rocket launchers were previously installed, these trucks are extremely fast and can cross complex terrain.

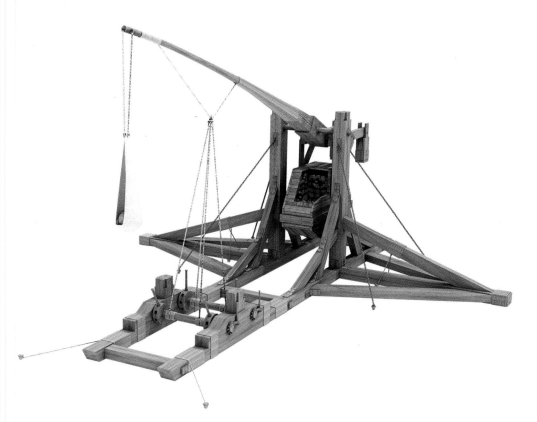

U

V

W

X

Y

Z

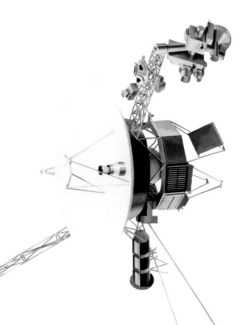